This book was made possible through the generous support of Terra Moretti.

Attilio Scienza

The Art of Wine Storytelling

Mamma Jumbo Shrimp

The Art of Wine Storytelling

Author
Attilio Scienza

Foreword
Stevie Kim

Translation
Richard Hough

Editorial Team
Giulia Stocchetti, Michela Aru

Graphics Team
Sara Zambon, Andrea Darra

Prima edizione: *Marzo 2025*

Via Adriano Cristofoli, 16
37138 Verona
https://mammajumboshrimp.com/

ISBN: 978-88-946863-0-2

Sommario

Foreword by Stevie Kim

It's no accident that this book is being published to coincide with Professor Attilio Scienza's 80th birthday. A neat little tribute? Perhaps. A blatant excuse for another round of celebrations? Most definitely. Not that Scienza needs an excuse—he has more energy, dynamism, and intellectual firepower than people half his age, which is frankly quite unfair. I can personally attest to how physically (and mentally) exhausting the wine world can be, yet while many in his position might be enjoying a well-earned retirement, sipping a fine vintage in some sun-dappled vineyard, Scienza remains one of the hardest-working, most in-demand voices in Italian wine. The man is like a walking, talking, lecturing, and occasionally ranting encyclopedia—if encyclopedias had a penchant for Aristotelian digressions and Homeric analogies.

And while he may not have a social media account, let's be clear—he never snubs it. In fact, I've personally volunteered him (more than once) to participate in a TikTok trend dance. No problem whatsoever. This is a man who will fearlessly take on biodiversity, sedimentology, and yes, even the occasional viral challenge—reflecting his openness to everything, even when it's completely foreign to him. It's this same openness that keeps him not only relevant but constantly evolving in an industry where tradition often resists change.

So think of this book as a reciprocal birthday gift: his gift to us and our gift to him. Although, let's be honest, it's mostly his gift to us.

Of course, Scienza's real gift isn't just his relentless work ethic (which is exhausting just to witness) but his storytelling. As listeners of "Everybody Needs a Bit of Scienza" an Italian Wine Podcast series dedicated to Il Professore know, he can wax lyrical on subjects ranging from biodiversity and climate change to mythology, tectonics, and sedimentology. He'll casually drop insights that could take months to unravel, delivered in a way that suggests he's merely scratching the surface. And while it may seem like he's speaking off the cuff, don't be fooled—every answer, every lecture, every unexpected tangent (and there are many) is rooted in a lifetime of study and an insatiable curiosity. I won't pretend I always follow everything he says—who does?—but I can say that his enthusiasm is nothing short of contagious.

And let's talk about that career choice. Scienza could have been a leading scientist in any number of fields, but, in his infinite wisdom, he chose wine. And lucky for us, he did. His impact has been so profound that I often joke that if he spoke English, he'd be more famous than Jancis Robinson. But, alas, he doesn't. Which is precisely why we're publishing this book in a bilingual format—to give his ideas the wider audience they deserve. If you've ever seen him at Vinitaly, swarmed by admirers and juggling more events than seems humanly possible (with an itinerary that makes mine look leisurely—and that's saying something), you'll understand just how revered he is.

For those fluent in Italian, you get Scienza in his raw, unfiltered form—about as close to the horse's mouth as we dare to publish. But let's be honest, Scienza doesn't just speak or write in Italian. He speaks and writes in Scienza language. Yes, technically it's Italian, but it's Scienza's Italian. So much so that when we translate his books, it's often a two-step process: first, we translate from Scienza to Italian, and then from Italian to English. Even his PowerPoint slides require a bit of deciphering before they can be shared with a broader audience. Thankfully, we have what we affectionately call "Scienza whisperers" on our team—people specially trained in untangling his uniquely complex phrasing. And for this book, Richard lent a hand in that delicate process.

Because that's what makes Scienza's writing so, well… Scienza. It's colorful. It's unpredictable. It's occasionally maddening. It poses more questions than it answers. It scratches intellectual itches you didn't even know you had. It meanders, it provokes, and just when you think you're following along nicely, it takes an unexpected left turn into ancient philosophy or 19th-century viticultural crises. Which is precisely why it's so compelling. Perhaps it's time to brush up on your Homeric epics. Or revisit your high school biology notes. But more importantly, it's time to rethink how we tell the story of wine—beyond the usual tired clichés.

The working title for this book was Roots and Reason: The Evolution of Wine Storytelling. It was a title we were rather pleased with—a nod to both the deep historical roots of wine and the reasoning that helps us navigate its future. It also followed neatly from our last Scienza publication, Vine and Prejudice: Fake Science and the Search for the Perfect Grape. But in the end, we went with The Art of Wine Storytelling—a title that, while perhaps less literary, has the advantage of being immediately clear to an Italian audience. It does, however, lack the rhetorical flourish of the Jane Austen-inspired alternative. Although, to be fair, while Austen may not have read Aristotle, Aristotelian concepts are all over her writing—just as they are in Scienza's work.

From prehistoric cave paintings to the mythical voyages of Ulysses, wine has always been a central character in the human story.

Scienza takes those ancient storytelling techniques—the heroes of Homer, the rhetoric of Aristotle—and applies them to more recent wine history, from the battle against Phylloxera to the rise of the Super Tuscans. But, more importantly, he shows how these same techniques can be used by wineries today to tell their own stories in a way that resonates with modern consumers.

And that is why we are once again delighted to share the insights, expertise, and boundless enthusiasm of this one-of-a-kind storyteller.

Happy Birthday, Professor Scienza! May you continue to inspire, provoke, and occasionally infuriate us for many years to come.

Professore ti vogliamo bene. Tanto.

Stevie Kim

Storytelling: The Factory of Stories

Lisa Cron, in her book *Story or Die* (2021), argues that humans are not born to read, that reading is not a natural act, but that we are born to communicate and tell stories. Our brains synchronize with the protagonist of a story through the narrator, activating the areas responsible for processing emotions, allowing us to empathize and put ourselves in someone else's shoes.

Storytelling, which originated almost simultaneously with humanity, has always been a critical tool for sharing social values. Its origins can be traced back to the first prehistoric cave paintings. While the great ancient narratives told universal myths that shaped human history, modern storytelling works in the opposite way. Instead of recounting past experiences, it describes reality through constructed tales, shaping emotions and creating narrative frameworks that allow people to connect with stories.

From Marco Polo's travel diaries to the adventures of Robert Louis Stevenson, the concept of narration permeates every field of knowledge, from political science to anthropology, medical research to molecular biology, persuading, expanding understanding and sharing experiences. The so-called narrative turn in social sciences that took place in the mid-1990s coincided with the explosion of the internet and advancements in information and communication technology.

The words of the Israeli writer Nomi Eve define the essence of storytelling, where even modern stories can be understood through the lens of legend:

"The word legend comes from the Latin 'legere,' which means 'to read.' The word fiction comes from the Latin 'fingere,' which means 'to form.' From fingere we also get the word fingers. We form things with our fingers. The word history comes from the

Greek 'istor,' which means 'to learn' or 'to know.' I believe in original etymology. I believe that fiction is formed truth. I believe that history is a way of knowing all of this. I believe that legend is how we read between the lines."

This myth and imagination intertwine in Greek tradition, as wine becomes a medium of experience, echoing the voyages of Odysseus and the Greek settlers who brought viticulture to the West.

The Power of a Story

Having a compelling story to tell is vital and strategic and has enormous potential for brand differentiation and recognition. The essential question is: "What is the essence of the story to be told?"

Every convincing story originates from an inspired narrator. For wine producers, the mission must transcend mere profit to communicate a deeper purpose to their audience. Origins, aspirations, challenges and successes form the foundation of such narratives.

Defining your mission is the first step in crafting a story that resonates. But this requires technique. How do you distinguish between a generic mission, a formally correct one, and one that touches hearts and demands action?

Creating a meaningful narrative requires delving into the intrinsic significance of a wine's genesis and production, the effort that has gone into it and why it should be preferred, even desired. A wine's foundational identity lies in its story, not just in an eye-catching label.

From here, the journey begins, setting out complex stories that touch the realities of our world. Philosophy or aesthetics are not enough to ennoble a wine. It is necessary to go in search of a deeper and more intrinsic understanding of its genesis and its production.

Anyone who has dealt with science has had to deal with the theme of storytelling and that of Greek culture. Narration is a process inherent in our biological memory that is based on certain *topoi* or narrative schemes. These patterns are a sort of biographical propeller on which our narratives flow and are based on certain key elements. The hero in search of himself, the quest, the antagonist who opposes the hero, the conflict between the hero and the antagonist, the treasure, the event to be discovered to complete the quest, the adveristy that the hero must overcome, the magical objects and people who help the hero, the final prize on completion of the quest.

Each of us feels a bit like a hero. Our daily lives reflect the hero's script, with each of us searching for some kind of treasure at the end.

Homer, the father of Western storytelling, develops the theme of the hero who goes in search of himself, lives adventurous and dramatic experiences, is helped and returns to his native land, to his dearest loved ones.

In summary, the storytelling of wine and its production should grab the attention of listeners / readers, create new meanings, fix related symbols and impressions in the imagination, provide a better understanding of the qualitative aspects of a wine, and provide context and relevance to the "company". Effective storytelling creates empathy and increased user and product engagement, building brand and logo loyalty.

The Role of the Consumer

Consumers think, remember, and decide based on stories. Storytelling can be considered a "factory of stories". As Julius Caesar noted in *De Bello Gallico*: "Men willingly believe what they wish to be true."

Storytelling is a device for capturing attention by means of the story, the plot and the narrative tension. It doesn't attract attention like the logo does, but it builds audience loyalty thanks to its narrative flow. The human brain is better able to synthesize information when it is communicated in narrative form. From schemes, figures and lists, stories are created using narrative techniques developed from fiction, comics and cinema. The immaterial cues for a narrative are numerous: the territorial appeal, the socializing function, the identifying function, the projection of the self, the adherence to a lifestyle / consumption, the identity of the producer, the myth / quality.

The material cues are a little easier to interpret: the taste-olfactory profile, the nutraceutical function, the healthiness, the sensory impact, the physiological impact, the psychological impact, the materialization of the packaging, the brand guarantee, the price. All these elements are on the one hand the physical substance of a bottle and on the other the spirit of the bottle. It is difficult to build a story that resonates only through material elements.

The story of wine therefore starts from the story of the territory. The territory has history, anthropology, geography, soil, climate. These are all fundamental elements to tell a story because they provide a sense of place. We cannot invent a history of a wine without being able to place it in a semblance of reality. Around this we can build whatever we want. The most effective language is that of the senses. The territory can be explained not only as a physical description but also in terms of how our senses interact with it and the impact it has.

Of course, the starting point is the product, we cannot start from a territory without a quality product. The profile of the producer, the style of wine consumption, the genesis of the myth are also very important, and on the myth, we can build the story of the hero. No other product category has been able to develop its own poetics as refined, complex and detailed as that of wine production. Enology is thus transformed into "enolosophy", and wine can help to take us from a rationality that is too soberly humane to the elevated status of a sacred symbol. It is important, however, that entrepreneurs, wine communicators, those involved in retail sales do not fall into the trap of the rhetorical component of the rich repertoire of fairy tales that have contributed to creating and developing some nonconformist reflections about the possibility of adapting wine to the tastes of consumers, especially those of tomorrow.

We need serious communication, rich in technical content, made with a simplified language, which leaves the consumer freedom to choose, and not just emotionally. A few guiding words are enough: landscape, territory, culture, environment.

A story is needed

Wine is the result of the coming together of vines and winemaking techniques with distant and mysterious origins, which only with the progress of tradition has found a perfect synthesis in the wines we know. It is a fusion that satisfies, more than any other agri-food product, the desires of eclecticism and syncretism of the modern consumer.

Eclectic is the one who moves with ease combining different styles, syncretic because he manages to blend them harmoniously, creating a seductive synthesis.

Today it is increasingly difficult to discern the influence of aesthesic taste (the connection with the sensitive properties of a wine or food) from the relevance of aesthetic taste (the ability to discern beauty or imperfection). For a wine it becomes essential to identify and isolate the different from the undifferentiated and this can be a very gratifying process for a wine lover, capable of producing a sort of euphoria, in a context that is prevalently dysphoric, which tends to flatten, homologate. This directs the taste of each of us towards the demand for a growing variety of types of wine, which are often distinguishable only by the labels.

In the past, "good taste" played an important part, the exercise and application of those rules that characterize the hegemony of one social group over others (the bourgeoisie of the late 1700s, for example) or the adoption of the "single thought" of some tasters capable of cataloguing wines according to easy-to-define sensory profiles (Bordeaux wines, for example). The recent history of wine consumption shows us a growing series of anomalies that have had the effect of including or excluding certain types of wine or even entire territories, in the restricted circles of power of a few brands. Globalization, of which everyone now perceives not only the semantic meaning of the word, but above all its impact, influences not only events on a global scale, but also our daily life, including our relationship with wine. In the world of wine, globalization has brought obvious advantages, especially at a mercantile level, expanding the consumption of wine among peoples who barely even knew of its existence, but the disadvantages should not be underestimated. The growing simplification of the kind of wines that are produced today all over the world, often undistinguishable, regardless of grape variety and terroir, strongly conditioned by the "single thought" precise sensory models of "perfect" wines, but without a soul.

The reaction to such wines has manifested itself in alternative forms of winemaking and viticulture, such as the return to native vines, vinification in amphorae and biodynamic viticulture, but the fate of many wines produced from native vines, which are characterized by their original sensory profiles, differ too much from the descriptive models adopted by those who decide the ratings in the guides. The question remains, however, of undoubted importance for those who are involved in the marketing of wine, where the problem arises for those who produce it or communicate it to highlight the most important characteristics. Sometimes the consumer perceives and is attracted more by the linguistic subtleties with which the wine is described by experts who are often not influential enough to convince consumers.

It is the places of production, the territories and the welcoming cellars, that are increasingly able to give the evocative impact of the product to the imagination of the wine lover, but it is an elitist occupation reserved for a few.

The fate of wine, whether you like it or not, is in the hands of the big retailers, who are operating a substantial over-investment in wine, represented by a good price, as wide a range of products as possible and the evocative setting of the "enoteca". This information is certainly not new, at least for insiders, but it highlights a deficit of marketing strategies that fails to grasp the changes taking place in an increasingly fluid society, where too much information circulates and where it becomes difficult for the recipient to extract meaning of real value to them. We are also witnessing a growing process of de-semanticization of language and attrition of symbolic processes. Hopefully there can be a renewal in the relationship between producer and consumer capable of transforming threats into opportunities.

Horace's expression of "mixing the useful with the pleasurable" (*miscere utile dulci*) is the perfect sentiment in this regard. In his *Ars Poetica*, he advocated the use of narration to inspire the reader, carefully combining fun with a little learning.

The planet is changing, society is evolving, and wine is not exempt from those phenomena that affect the world of consumption. Wine, due to its symbolic and societal status, is affected more than other economic sectors by these changes to our ways of life.

We are witnessing a return of the social and this has important effects on the market. We have gone from a consumer who based their choices on the impulses of the ego to differentiate themself from others, to stand out from the crowd, to a consumer driven by phenomena of emulation, fusion. For many, drinking wine now means searching for roots, returning to forgotten places, celebrating a memory, renewing a fraternity or reliving a common experience.

The verb "excel", which derives from the Latin *excellĕre*, which means "to push out", does not signify a superior level in the same field, but to enter a different field. To move from good quality to excellent, it is not enough to explain the wine, but it is also necessary to communicate the intentions and commitment of those who are involved in producing it, their values and vision. It is necessary to present not only the product but also the producer and the inspiration and methodology behind the product.

The quality of a wine, therefore, based on its sensorial characteristics, can be sufficient or good, but to be excellent it must have additional intangible values that do not reside in the wine itself, but in its spirit, in the humanity of those who produced it, and in the relationship with of those who drink it. The first intangible element that the consumer perceives is the credibility and reliability of the producer.

Emotions, symbols and communication of wine

We are prey to our emotions, but without emotions we wouldn't be able to meaningfully participate in the events of our lives and interact with others. Without emotions we would be unable to appreciate information that comes to us from art, society and science. Emotions are linked to the organs of sense, sight, hearing, touch and taste. Would it even be possible to appreciate a work of art or a piece of music or taste a good wine without emotions? Are emotions the expression of pure and simple mechanisms of behavioral adaptation or are they the result of "constructions", that is, a synthesis that integrates and amalgamates sensory sensations, memory traces, micro-behaviors, textual routines, expectations?

Emotion is not only based on physical-psychic perceptions but requires connections with the imagination, understood as the result of the compatibility between the mind and the external world, the so-called "common sense". In 1995 Daniel Goleman wrote a book entitled *Emotional Intelligence*, in which he described the subject matter as a tool that allows us to read emotions and weave relationships, to get to know ourselves more deeply and evaluate the reactions of those we interact with. The seat of emotions is in our brain, in our temporal lobe, in a teardrop-shaped structure called the amygdala. It evaluates the stimuli that come from the outside world: anguish, curiosity, desire, empathy, nostalgia, surprise.

Emotions can also shape our perceptions. The wines of Etna, for example, are celebrated for their quality, but how much of this relates to the observation that its soils are of volcanic origin? How much is our perception of the beauty of the Bolgheri region enhanced by the fact that Sassicaia is produced there? The emotion of taste is therefore a product of the exercise of the imagination. In the managerial code, in the "experience economy", the production

of surplus value is closely linked to the so-called "emotional experience", and to the sphere of the imaginary through the production of artifacts with a high symbolic component (the brand).
Empathy is the ability to recognize and identify with what another person is thinking or feeling and to react with a similar emotional state; to feel an emotion together. Mirror neurons, the subject of research by Giacomo Rizzolatti, neurophysiologist and professor emeritus at the University of Parma, are a particular class of neurons present in different areas of the motor cortex of the brain that are activated when a human being performs an action and at the same time this action also occurs in the observer of the action itself. They are, therefore, important to imitation learning. They can also be defined as the neurons of empathy and can have a significant impact on relationships between people.
Smile or kindness education, which allows this demeanor to be transmitted to others, shows that the smile is transferred through mirror neurons: almost a parallel brain dedicated to social relationships, emotions and actions of various kinds, close to that part of the brain that deals with rational thought. One theory even links the dysfunction of the mirror system with autism.
The impact of mirror neurons is also apparent in the worlds of theatre and sport. For example, a basketball player guesses from how his opponent's hands are placed if the ball will go into the basket. In collective social manifestations, especially where very deep feelings are evoked, mirror neurons are activated, and the viewer participates as if he were present in those scenes. The invention of Cinématographe (the motion-picture camera and projector) by the Lumiere brothers transformed cinema from an individual phenomenon, for individual experiences, into an event shared by wider audience.
So too, wine has become a social phenomenon, from a culinary

experience to a shared emotional one. Consumer psychology has demonstrated the importance of the public as an imaginary element in the mind of the wine producer, an audience that can also be made up of a single person, who lets himself be carried away by the suggestions, the sensory stimuli of the wine.
So, what is the contribution of mirror neurons in creating the climate of empathy between those present in a tasting and, therefore, ultimately in the positive perception of the sensory characteristics of a wine? In the collective behaviors of a modern tasting, we can see the ancient symbolic ritual contained in the Greek symposium and in the Christian mass and which can be defined as the "principle of incorporation".
Today's consumer unconsciously poses two important and substantially new problems: that relating to the destruction of food or wine in its material form and that of transmutation in the tissues of his body. Two aspects inherent in the act of nourishment are resolved in the classical world through Dionysian rites. The god who dies twice, with the crushing of the grapes and the rebirth due to fermentation in the act of drinking, and in Christian mysticism with the death and resurrection of Christ. The attempt, therefore, after the Cartesian separation of the spirit from the body to reunite them by identifying the outside of the body, the food, with the inside, the vital force that identifies the spirit that reunites it with the earth.
Carl Jung in the phenomenology of the mass recognizes in "eating the god", the meaning of *syn-ballein* (from which the word "symbol" is derived), being together, the alchemy to transform the raw material of wine, into a new psychical reality, the submission of biological reality to the spiritual need that strengthens the social bond.
In the distant past, wine was at the center of the various manifesta-

tions of Western societies, such as celebrations for births, weddings, rites of passage up to the burial of the dead. Now a phenomenon of de-symbolization is taking place, which testifies to the progressive distancing of modern man from the suggestions of myth.
What are the consequences on the sociology of wine consumption? In the Greek symposium as in the shared consumption of wine nowadays, where a wine is accompanied by interesting conversation shared around the table, it is difficult for the evocative aura to lead to drunkenness. It is the drinking of young people who want to get drunk quickly, not so much drinking on an empty stomach but drinking with an empty brain.
Like it or not, we must return to the religious significance of wine. Not so much in terms of faith, but in the sense of belonging, by which we understand the profane customs and ceremonies that legitimize a society and link it to a territory. In wine we find the scent of a landscape, of mother earth. Those who choose not to drink wine send an important message: they do not belong to this land.
It is hoped that in this period of crisis, as happened at the time of the methanol scandal, there can be a renewal capable of transforming threats into opportunities. Put in brutal terms, the world of wine must find a convincing answer to the radical consumer question "why should I drink (your) wine?"
In recent years we have witnessed the proliferation of wines, mainly managed by marketing men, with the workers of the vineyard in the background, their role reduced to the bare minimum. A triumph of image over substance. An attempt was made to create an almost magical expectation of enjoyment and pleasure with no expense spared on special effects. Then, at the moment of truth, the encounter between the much-vaunted wine and the taste buds of the consumer, there was often a significant gap between expectation and reality, both sensory and emotional. This sequence of

seduction, illusion and subsequent delusion, leads to a crisis of credibility and trust and translates into a continuous slump. A bit like the syndrome of the seduced and the betrayed lover, in this case taking refuge in other categories of wines.

It is desirable, therefore, to pursue a more sober style of communication which creates direct channels of engagement between the producer and the consumer, enhances the connection with the product itself, promotes greater information transparency, and resonates with the recipient. This can be achieved by conveying information which emphasizes the passion and contribution of the winemaker, as well the common bonds and shared values between the producer and consumer. In short, communication that should abandon itself to the pleasure of telling and re-telling stories.

Communicating wine: an art for the few

We are witnessing a great change in the approach to judging a wine. As in cinema or fiction, where criticism can represent the end point for a book or a film, criticism can also represent the final point for a wine. Comments provided an interesting new perspective. They opened up worlds and showed new and sometimes surprising ideas. There was also the judgment of ordinary people who drank a wine or read a book and talked with friends. With the existence of social media networks everything has changed. First, because the democracy of the web does not easily tolerate the authoritativeness of the critic (to be honest often not very authoritative). And then because common sense cannot be summarized in a "like" or "don't like". We read opinions, judgments, expressions borrowed from almost everywhere, we try to prove that we are better than others, but often we lack the necessary experience and knowledge.

One result, however, has been obtained. Eliminating the unique thought of opinion leaders who have codified the canons of quality, often not for the merits of the wines but for the benefit of their members.

When presenting wines, we must avoid using emphatic and redundant language from a lexical point of view, full of rhetorical and syntactic figures, full of subservient and nominal phrases. Avoid proposing the false distinction between a technical evaluation and one of pure pleasure. Avoid transforming technical sheets into tributes or tools that claim to do everything: describe, analyze, classify and reward.

Stories are fueled by emotion. Our lives are lived on emotions. The most engaging stories are those that enhance vulnerability by making us identify with the protagonist, that arouse a more intense emotional response. A story captures our attention, drags us into its world, into a reality that has replaced our own.

For Aristotle, when a story is created, consisting of a beginning, a middle and an end, it develops into the structure of the "plot". The plot is what is visible and what conquers us, but it is not so. What makes the plot so compelling is the internal effect, in the mind of the protagonist, that forces him to assume a new awareness, thanks to which he will be able to overcome the obstacle that prevented him from solving the problem. For Nietzsche there are no facts, only interpretations. Communication specialists have discovered that the human brain is capable of synthesizing information at a multisensory level when it is presented in narrative form.

Storytelling is a tool for capturing attention by means of history and narrative tension. But language must develop through a discourse of truth, which requires the fulfilment of certain conditions that Michel Foucault (2016) illustrated in his seminar on parrhesia, (speaking frankly, speaking truthfully, telling the truth), referring to the loss of democracy in Athens in the fourth-fifth centuries BC. Foucault's rectangle is a concept that allows us to ideally synchronize the formal conditions (fair competition), de facto (the power of monopolies), truth (reasonable *logos*) and moral (the courage of struggle). The theme is very current for the relationships that are established between social media and all kinds of propaganda, not just political. From tools of freedom and liberation of information, sometimes also through AI, social media seem to have transformed into a disturbing tool of disinformation and mass manipulation through the now uncontrollable spread of fake news.

Wine and society: an evolving relationship

Market research reveals that among Italians who drink wine regularly (approximately 60% of the total population), only 2% consider themselves experts, while 20% feel they have an acceptable level of knowledge. A trend has also emerged which sees wine paired with food in increasingly unconventional ways. The ideal wine, particularly for younger consumers and those from new markets, appears to diverge significantly from the current offerings shaped by producers and tradition.

Women tend to prefer spiced, aromatic, and floral wines, showing more openness to novelty (including ethnic cuisines), leaving tannic and structured wines to men. However, men increasingly lean toward lighter, less alcoholic wines. A progressive disconnect emerges between the experiences and imaginations of consumers and the rigid thinking of producers. This highlights the need for a revision in how wine is communicated today.

Historically rooted in Mediterranean culture, wine's evolution reflects societal changes. For centuries, low-quality wine was a staple dietary component for the poorest social classes. Until the 1960s and 70s, wine was predominantly "homemade", purchased in bulk, and locally sourced. With Italian society's break from its agrarian roots, wine consumption mirrored other luxury products. The masses began emulating the habits of privileged classes, claiming cultural ownership, establishing clichés and, most importantly, engaging in conversation about wine.

The habit of drinking alcoholic beverages does not happen spontaneously, but is shaped by family and social group behaviors. Young people today rarely drink wine at home and tend to frequent social spaces where wine plays a limited role. Where, then, will they learn the sensory vocabulary crucial to understanding wine? For younger generations, wine suffers from significant voids on a symbolic level, despite efforts to address these through various educational initiatives. Wine culture has become a bookish top-down media driven subject led by experts who are often not sufficiently persuasive to convince consumers.

The world of wine is steeped in symbols, some of which we consciously recognize, while others remain deeply rooted in our psyche. The myths tied to wine bear witness to our rituals. For a ritual to be effective, participants must share the same myth, aligning behaviors within a community. When wine transcends group habits and metaphorically enters the global stage, its symbolic language loses its mythological meaning, shifting towards new ritual expressions and forms of communication. For example, are we sure that the language we use to communicate wine to Western peoples and which is based on expressions of myth inherited from the Greek world, is also effective for the peoples of the Far East who have developed in their culture, myths and symbols so different from ours? What relationship exists between a supposed universality of myth and tradition? The answer lies in the origin of symbolic language, which does not belong to a universal knowledge that can be easily shared by all civilizations, but which is instead the expression of regional knowledge, conveyed by that mythological arbitrariness, where anthropologists call arbitrariness the divinity that represents the original, the beginning of history. And what preserves the sense of history if not tradition?

But the risk is that the ritual becomes a ceremonial formality that in its reiteration is void of content if it is not constantly integrated with those elements of "ritual augmentation" that consist of the technical-scientific insights inherent in the processes of formation of sensory profiles (geological origin of the soil, characteristics of the climate and grape variety, etc.) or in the stages of wine production. Research and knowledge to grow the culture of wine and attention to the raw material, also from a holistic perspective of biodiversity, the conservation of centuries-old traditions, the anthropological and sociological aspects of wine production.

The New Narrative Era

Since the late 20th century, brand storytelling in the United States and Europe has been transformed through media industry mechanisms and globalized capitalism, becoming a powerful tool for shaping consumer and citizen opinions. Storytelling, once reserved for children through cartoons and grandmotherly tales, evolved into the "narrative turn," (a new narrative era) giving rise to narratology. Marketing technologies have shifted from product-centric approaches to logo-centric strategies, and eventually to stories. A brand's success now lies in the narrative that communicates it.

The crisis of brands is caused by the elusive, unpredictable, paradoxical behavior of consumers, who are no longer as fascinated by the brand. Brands have started speaking, and when they do, consumers listen carefully. What's in a name? (Shakespeare's *Romeo and Juliet*). In 1996, Pakistani children hunched over soccer balls with the word Nike went around the world: inside the brand there were stories of misery and hunger. The brand Levi Strauss, for example, became synonymous with the working-class hero archetype, selling jeans by telling the stories of miners. This narrative universe, or "dream society," fosters shared visions and connects people globally. Consumption thus becomes a planetary communion of symbols, mirroring the Catholic Church's concept of communion among saints. By consuming mythical symbols (e.g. wine), we internalize the most potent myth of all — heroes (e.g. Star Wars).

Storytelling is the science that translates and promotes things (real or imaginary) into words, images, sounds and real perceptions and for this reason makes them true, legitimized to exist. It is an approach common to many sciences, from anthropology to

genetics, from neuroscience to geology, it becomes a managerial discipline, an indispensable tool to be listened to and chosen.
Pop culture is obsessed with storytelling. The men's fashion week that takes place in January each year in Milan and then especially in Paris in June, is a riot of *storytelling*. The richer in *pathos*, the more effective the narrative. Fashion shows have now reached new heights of theatricalization: clothes are only one of the elements, sometimes not even the main one, of complex staging that stimulates the senses.
In general, in the new storytelling it is the urban environment with its critical issues that prevails. What matters, these days, is only the image. Understanding the aesthetic value and the design path of the collections requires an awareness of the representation in front of you: clothes must be seen as masks, as costumes, not as simple items of clothing. The latent danger of a disconnect between what is seen on the catwalk and the actual reality of the sale has finally materialized. Fiction prevails, because the clothes on the catwalks are only a lure to social networks. In the store, you will find something else. This dichotomy has always existed, catwalks and newspapers require high-impact gimmicks. Clothes to wear, on the other hand, must also serve a practical purpose. Has this dichotomy also arrived in the world of wine?

From the Odyssey to digital storytelling

The growing distrust of classic advertising, the so-called brand communication, is the consequence of the advent of the internet which has triggered a profound change, together with social media, represented by digital storytelling. This change is characterized by the transition from a so-called asymmetrical communication implemented by brands through massive investments with repeated messages in traditional media, to narrative sharing in which the story suggested by the brand is integrated and reinvented by the public, through the creation of spaces for dialogue where they can express themselves freely and share experiences. Blogs can perform this function well. Such content, which is intended to communicate an idea, is represented by blog posts, podcasts, photos, videos, comments on YouTube. The story that takes shape, travels on the web and is enriched with unexpected details, developing interactive stories with a multitude of users.

Digital storytelling uses digital tools to build multimedia stories with a strong emotional impact. Digital stories harness their expressive potential in the use of images, films, music, people's stories. It is mainly used in teaching, training, learning, to share brand values, for marketing or as a communication tool within companies. The case studies are numerous, even if there is a lack of trainers and storytelling experts, largely due to an inadequate culture in audiovisual and multimedia languages. By means of digital technology and YouTube, anyone who owns a camera and has a story to tell can find their audience. The phenomenon of online video sharing is destined to represent a communication channel of increasing importance in the future. Blogs are an example of storytelling that can help to overcome barriers of impersonal corporate communication and interact with the various

stakeholders, internal and external, in more direct and innovative ways.
The use for narrative contexts is the best tool to capture the attention of viewers who are increasingly distracted by infinite digital connections. But is such an approach really new?
The success of Homer's *Odyssey* is largely linked to his ability to narrate the facts according to a plot that served the Euboean navigators for the conquest of the West. In fact, it can rightly be considered a travel manual used to exorcise the fears of the crews of ships heading towards the unknown. The places of the archaeological finds of Euboean skyphoi (a type of drinking cup used in ancient Greece) coincide not by chance with the places frequented by Ulysses.
Our perception of the history of the ancient Greek sailors struggles to distinguish truth from falsehood, reality from fiction. The superiority of Greek literature consists precisely in its relationship with reality, with space, with the idea of frontier and conquest, through the references to places of history and the myth of the Mediterranean. For a long time, the Mediterranean coast represented more than a destination celebrated by stories, a "narrative horizon" towards which the peoples of the Near East headed, a place where anything was possible, and which offered everyone the opportunity to start a new life. At the same time, it was both nation and narration.

The structure of organized storytelling

The main elements of storytelling are represented by the story (which focuses on the "what") and the plot (which focuses on the "how"). In the story, together with the actions and events, the characters move in a defined environment. The discourse is instead articulated on the structures chosen for transmission and in the manifestation in the media. Normally the narrative is organized on two models, the so-called all-encompassing one, which develops on sagas (epics and legends) and stories (short stories, podcasts and novels); and the fragmentary one, used mainly by the media (gossip, news and rumors). They are two different ways of communicating, in the second case it is a very rapid and non-exhaustive form of storytelling, which serves to create curiosity, while with short stories or novels the narration offers stimuli and arguments that can be developed with other techniques, such as cinematography or theater.

A performance of wine storytelling must bring together four key elements: the story; its setting; the subliminal effect of vocabulary and metaphors; and, fundamental in a novel, a stimulus or narrative tension that invites the reader to turn the next page. On the internet and social media, narrative contagion is increasingly important.

In the case of the narrative related to wine, an interdisciplinary approach is necessary to involve all the elements that are the basis of its production. The narrative macro-areas include humanistic aspects (historical, literary, archaeological, anthropological content) and scientific skills, such as those related to pedo-climate and grape variety. This is a challenge because it is difficult to find all these skills in the people who write about and communicate wine through *storytelling*.

As in all self-respecting stories, a narrator is also needed.
What is role of the narrator?
Two aspects are fundamental: the allocation of the story to a literary genre (thriller, mystery, romantic, non-fiction-scientific, fantasy); and the choice of plot (epic, tragedy, melodrama, comedy), bringing together the rhythm of the story with the unfolding events.

Venice, Malvasia and the birth of modern storytelling

Venice and the trading of Malvasia wine is a good example of storytelling because it uses an anticipatory communication strategy that goes beyond the canons of the classic narrative previously employed by the Mediterranean wine trade, relying on a very simple strategy that was based on two tools that are still current, the brand (the name Malvasia) and notoriety (an exclusive wine for the European courts).

Venice sensed that it was necessary to transform wine from a food item into a cult object, an icon. If the Greeks had linked wine to Dionysius for ritual use in the symposium, Venice loaded the wines of the eastern Mediterranean with secular values, transforming them into a fashion, a remedy for the body and spirit in a Europe afflicted by the amine and Black Death that followed the Little Ice Age.

Venice did not use the Greek symbolism of death and resurrection (from grapes to wine), nor that of the mystical blood of the Christian rite, but instead relied on the more modern phenomenon of the status symbol, closer to the models of imperial Rome, of the place of origin and of rarity. Adopting *ante litteram* the most effective methods of contemporary communication, linking those wines to an elitist consumption, giving them to the rulers of the time, offering them during official banquets where they would be tasted by diplomatic corps from all over Europe, Venice applied high customs tariffs to these wines to limit their consumption only to the wealthy classes, while valuing the sense of origin and the allure of distant territories full of mystery, such as those under Byzantine control. Using the name of the place to market it, an absolute novelty in an era in which wines were designated with very generic indications, "*de plano, de monte*" or grape variety, it offered consumers a type of wine, sweet and aromatic, then unknown in the West.

Synesthesia and Proust Syndrome

Synesthesia, in the strict sense, is a neurological phenomenon in which the stimulation of one sense automatically causes a response in another. Normally, in sensory perception - be it visual, auditory, olfactory or tactile - one modality prevails over the others. In the case of synesthesia, on the other hand, seeing a letter or a number spontaneously leads to recognizing a certain color; listening to a certain piece of music or the sound of a word simultaneously leads you to perceive a specific flavor. It is not an association of ideas, but rather a real sensory experience.

By extension, the concept of synesthesia is also borrowed to describe multisensory and evocative experiences in which different sensations intertwine, also evoking memories and emotions. In wine tasting, this multisensory dimension plays a fundamental role. The characteristics of the wine can activate deep connections with memory and individual experience, evoking images, sounds or emotions linked to moments lived.

Two significant literary testimonies can help us better understand this phenomenon. Gaston Bachelard, in his *Le droit de sûre* (1974), writes:

"This glass of clear, fresh, dry wine sums up my whole country life. People think I drink, but I remember."

Marcel Proust, in *In Search of Lost Time* (1913-1927), intertwines personal memories, historical facts, taste sensations and emotions through a diachronic and didactic approach to events. The episode in which, passing in front of the Swann house, Proust smells the madeleines is iconic and this reawakens in him childhood memories and the bond with his mother who made those sweets. The olfactory sensation becomes a key to open the trunk of memory.

When tasting a wine and presenting it to an attentive consumer, we should try to arouse, through the mechanism of the so-called "Proust syndrome", the memory of a pleasant sensation experienced in the past, evoking those sensory descriptors present in the multiplicity of foods, flowers or sounds that accompany all phases of our life.

Synesthesia and the minerality of wine: a taste of the past?

The term "minerality", a sensory descriptor increasingly used in the Anglo-Saxon press, associates this characteristic with high quality, originality and authenticity, especially for white wines produced in northern regions (e.g. Riesling). A sort of sense of place that evokes the relationship between the chemical (mineral) composition of the soil and the wine. For some it is a distinctive feature of some French wines such as Sancerre, Loire or Chablis and is linked to certain grape varieties such as Chardonnay, Sauvignon or Chenin Blanc. The Italian wines that best identify with this characteristic are Timorasso, Lugana, Greco di Tufo, Carricante dell'Etna and some aged Verdicchios. It is not easy to describe the sensory sensations it arouses as it has a multisensory and multimodal origin that combines smell, taste and trigeminal reactions. A saline minerality is distinguished from a bitter minerality. It is described as aromatic notes of flint, dry stone, musk, graphite, fresh oysters, smoky and kerosene, (or petroleum) in aged wines.

The research remains highly skeptical about the fanciful hypotheses of a relationship between the minerals present in the soil and the characteristics of the wine, and instead attributes these sensory notes to some sulfur compounds (thiols) present in reduced form, linked to the grape variety and the ripening conditions of the grapes. In fact, benzene-meta-ethyol gives an empyreumatic dimension to a wine very similar to the note of flint. Taste sensations are the result of heterosensory stimuli located in different parts of the mouth that transform chemical messages into electrical messages (transduction): acidity receptors, ion channels and membrane receptors.

All taste cells have receptors of these three categories that interact with each other, so it is therefore illusory to think that there are specific areas in the mouth for mineral flavor. The taste senses are neither analytical nor synthetic, only discriminatory and can make the difference, for example, between tartaric acid that evokes hardness and succinic acid that gives a salty or bitter sensation.

Why, then, is the attribute "minerality" so effective in communication? Perhaps because it recalls an almost physical relationship with a place. It satisfies that search for origins, the nostalgia for youthful memories, for holidays in the countryside. This accompanies that mutation of taste that rejects concentrated, woody wines, produced in a consumerist economic context with aggressive marketing strategies, and turns instead to more balanced products, coming from more natural terroirs that are able to reveal this minerality by virtue of a more eco-sustainable viticulture.

The perception of minerality then, beyond a certain balance between flavor and aromatic notes, is not the imperfect reflection of a soil but a mental representation that designates with a noun, a strong evocative power. The minerality of a wine can become a synesthetic stimulus because it brings to mind the naturalness of the place and the memory of childhood.

The magic of the story: the relevance of the myth

In the Italian wine-growing landscape it is still possible to encounter the reality of the myth. That's where the chthonian / primordial duality comes in. *Khthōn* means "earth" in Greek, a term found in the word autochthonous, which means "native to that land". In Greek mythology, *chthonian* deities (Gaia, Hades, Demeter, etc.) are distinguished from *primordial* deities (creators of the earth) such as Uranus, Zeus, Athena, etc.

What is their relationship with wine and the landscape? There is no doubt that a red wine, with structure, produced in a harsh landscape like Valtellina or sunny Calabria, is *chthonian*, a product of the earth, with deep roots preserved in the earth, in the depths of cellars and in amphorae made of terracotta. The ancient Greeks poured the first drops of wine on the ground in the ritual of the libation of the symposium, to appease the chthonian deities.

A sparkling wine, such as Prosecco for example, produced on gentle hills with romantic connotations of Valdobbiadene, has *primordial* characteristics, does not age in the cellar, is exposed to light with its clear bottles, is the wine of open-air parties, testifies to the beauty of the landscapes, and is certainly more suitable for modern socio-cultural practices.

There is also another duality, the Dionysian/Apollonian one, brought back to our time by Nietzsche. Dionysius is the god of wine, of excess, who drinks pure wine (for this reason called *Acratophorus*) as he does not pour the wine into the crater for dilution with water, as was the symposium custom. He identifies the wine as fleshy, strong, alcoholic, structured and once again the protagonist is the wild landscape, with wooded and rocky connotations. On the other hand, we have the wine of temperance, of spiritual elevation, of the Greek banquet, which was not a drunken gathering, but a moment of discussion, narration, song. Who then is the character who must be invoked? Without any doubt Apollo, *Phoebus*, the dispenser of light, who represents the pleasure of drinking as a moment for the spirit, not only for the flesh or for intoxication.

Tradition and reconstruction of memory

It is not the *genius loci*, the spirit of a place, that animates a wine-growing territory, but the *genius saeculi*, the spirit of the time, which allows us to narrate and think. The landscape is then a constructed space, where the spirit of place is not only to be found in the contemplative symbolism of the founding fathers and in the divinities that protect them but must also be evaluated with the eye of positivism where the work of the *homo faber* (the working man) admirably integrates the natural attributes of a place.

Neuromarketing, which is based on the development of cognitive sciences, uses storytelling to activate a process of synesthesia, through which a visual stimulus (the landscape and its landmarks) or a cultural evocation is associated with the sensory perception of the wine and the desire to reiterate the taste experience, providing the stimulus to think.

When we go in search of the meaning of the myth, to grasp its essence or its truth, what matters first and foremost is the collective memory with which a social group reconstructs its past by adapting it to the cultural contexts of the present, from which it can then shape the future. The example of the *Tabula Peutingeriana*, which sets out the road network of the Roman Empire, allows us to correspond the "places" identified on the map, with the wines (and vines) of the past.

In the meaning of myth, the paradigm of epic events is narrated as *fabula*, tradition must be used regardless of whether these wines are to be considered ancient or recent, original or reconstructed. The important thing is that the "mythical" tale is presented as a relic of tradition, but above all it is considered significant for the community to which it refers.

In the relationships between places and myth, the question is, above all, topographical. Places are still full of "signs" that in a *fabulosus* past find in their names the tool to document their origin in an indissoluble intertwining.
The *fabula* / place really becomes a "fairy tale" and a form of dialogue is established between place and story, where it is the wine that takes "meaning" from the ancient *fabulae*, which finds its foundation in the places where wild vines were domesticated and where their grapes were vinified for the first time. The myth is therefore wine.
The *topos* (theme) of the Trojan origins of Latin civilization through the protagonists of the *Homeric* and Virgilan *nostoi*, is not limited to the myth of the foundation of Rome and Naples but becomes the paradigmatic myth of Renaissance culture that identifies Greek heroes as the founders of many Italian cities. Padua, an Antenorian city, Mantua linked to Mantuo, son of the Theban Tiresias. Sulmona founded by Solimo, companion of Aeneas, Bari linked to the myth of Daedalus, brother of Minos.
Often the socialization of a place, of a city, according to structural anthropology, by connecting myth and language, is achieved through an object of recognition, a mountain shaped into a tower or a bridge crossing water, with the meaning of connecting, of uniting. What was the tool to read in that place the meaning offered by the natural world, and what is the basis of the process of transformation that progresses from nature towards culture, from the savage state to the civilized one?
Wine, as a result of the domestication of the wild vine, becomes a sort of milestone that marks time and renews memory. In Italy, wine-growing communities form an original culture in the agricultural world. Although strongly anchored in reality, winemakers have the ability to perpetually reinvent their past through an

emblematic and centuries-old product. Compared to the standardization of most sectors of contemporary society, winemakers occupy a cultural space halfway between rural and urban, tradition and innovation. In the relationship between the agricultural communities of Italy and the territory, the vineyard represents "the structuring element", a sort of ritual center, full of economic, political and symbolic significance, according to the French historian Fernand Braudel.

For anthropologists, earthly ritual centers have a material consistency (a wooden or stone artifact, a place full of symbolic meanings such as a vineyard) and constitute a real "anchoring function on the territory", representing a center of social aggregation with transcendental contents. Isn't the mosaic of Italian viticulture very different from this analysis, each a sort of ridge between different populations and languages, which carries with it the deepest, albeit less visible, signs of a defense of the "small homeland" and its independence, sometimes conflictual towards neighboring communities? Landscape therefore understood as a monument, a place of the spirit, from *monumentum*, a Latin word that encompasses the Indo-European root *men*, from which *mens*, intelligence, reason and *memini*, remember, think, reflect.

Looking at a geological map of Italy, the seemingly chaotic distribution of colors is striking, like some kind of complex mosaic. It depicts the families of geological matrices, the stratigraphic units, the types of rocks, and testifies to the great and varied richness of the soils that gave rise to it. Italy accounts for about 50% of the world's soil diversity. But that's not all. Italy is also the richest in viticultural biodiversity, with about 2,300 vines, of which 644 are registered in the National Register of Varieties and therefore approved for cultivation. If we cross-reference these data, we obtain a kaleidoscope of interactions of territories and

varieties, which make Italy a repository of wine and cultural experiences that has no equal in the world.
The study of viticultural landscapes, due to their strong character identity and their implications related to the quality of wine, is less and less oriented towards deterministic and economic aspects and more and more linked to cultural contents and social organization. Furthermore, since the landscape is a relationship between an observed spatial subject and an observing human subject, an unbearable interference is created from the point of view of the Cartesian analytical approach, which inspires our entire scientific method, since landscape science is to be understood as a science of complexity that must take into account different spheres of knowledge (human and material sciences, affectivity and rationality, physics of space and metaphysics of aesthetics).
In the collective imagination, aesthetic aspects and sensory characteristics often appear irreconcilable. What do they have in common? Historically, art and science have always been intertwined. Think of the Renaissance. Then their paths separated.
Why does art now matter less?
Why has science become less accessible?
The difficulty lies in making the so-called "two cultures" interact in a context in which more attention is paid to humanistic than to scientific culture, despite the fact that ours is a technological world.
Things are not miracles. Fire was not created by Prometheus. Science advances through the hard work of theory and practice, but culture is still myth. Beauty, a founding value, can become the element of fusion between diverse disciplinary fields. The methods of quantitative history are the new point of reference in overcoming an excessively anthropocentric vision of the geographical environment. The problems of industry are not separated from

those of agriculture and so the concept of agrosystem is developed, the study of which requires specific expertise in the field of ecology, in the geography of settlements, and in the natural sciences.

The analysis of context is key to understanding the relationships between ancient societies and the environment that hosted them, where micro-stories are the reactive containers where agricultural organizations, settlement structures and architectural forms interact in a complex puzzle of hydro-geomorphological, climatic and biological components (the vineyards). The retaining wall of a terrace or a vineyard, respectful of the morphology of a slope, is not an invention but the expansion of that which already exists.

The Importance of Rhetoric

Much of a university professor or communicator's work involves engaging with students and/or the public. Are they prepared for this? No, as Flavia Trupia aptly notes in her book about the importance of rhetoric and the power of words (*Viva la Retorica, Sempre! Il Superpotere della Parola*). Teaching and communication require specific preparation, which is often lacking. Before becoming a teacher, one must be a rhetorician. Those who can speak well sell products better, progress as students or professors, and captivate audiences during events. Rhetoric is a tool of persuasion.

Unfortunately, rhetoric is rarely taught in Italy though it is a formal subject in American universities. The word "rhetoric" often carries a negative connotation, evoking empty talk or time-wasting, sometimes attributed to politicians. A good rhetorician, however, is someone who can step out of themselves, focus on their audience, and achieve the desired impact through words. They observe their listeners, interpret their emotions, desires and fears, and adjust their message accordingly.

Perhaps the moment has come to reconsider the rhetoric of wine. Languages and narratives today use the same words very frequently, to homologate themselves to the point of flattening the perception of products that are very different in quality and reputation. In a profoundly changed world of consumption (generations, tastes, access and diffusion), work on language to diversify.

Rhetoric as a Tool for Winning Hearts

Public speakers are not immune from performance anxiety. One technique to overcome this is the *excusatio propter infirmitatem*, an explicit declaration of the speaker's own inadequacy. Sharing this discomfort with the audience turns a feeling of inferiority into an opportunity to appear relatable. The speaker acknowledges their unease in their role, creating a bond with the audience. This *excusatio* has been used by orators throughout history to avoid self-promotion, which can backfire. This humble approach, known as *captatio benevolentiae*, is a way to win the audience's goodwill.

The Five Steps of Rhetoric

The classic framework of rhetoric involves five steps:

Inventio: The invention or development of a thesis or idea to be articulated in the speech. The speaker identifies arguments to support the thesis, typically three or four. For example, when discussing native grape varieties, one might start by defining autochthony, proceed to the relationship between domesticated wild vines and cultivated ones, highlight contributions to local genetic heritage from migration, and conclude with strategies to combat genetic biodiversity erosion.

Dispositio: Structuring the argument. Start with an *excusatio*, present the thesis, illustrate the supporting arguments, and end with a final slogan or quote. Avoid overloading the audience with concepts or images, as this can dilute the message.

Elocutio: Refining the expression. This involves crafting the most appropriate linguistic style, often using metaphors to make the narrative engaging.

Memoria: Techniques to memorize the speech, ensuring the orator can deliver it confidently. While PowerPoint slides help structure presentations, a mental outline remains essential.

Actio: Gestures, intonations and non-verbal communication. Body language, tone, eye-contact and hand movements lend authority to the speech. Strategic silences can also maintain audience interest.

Among the rhetorical devices used in storytelling, metaphor is the undisputed queen. But, as Aristotle suggests, other means of persuasion are also required, represented by *ethos* (the speaker's credibility), *pathos* (the ability to leverage emotions) and *logos* (the ability to appeal to an audience's sense of logic or reason).

Another effective tool to be used for narration are the commonplaces of the mind, *the topoi* of Aristotelian rhetoric. They indicate a motif, a recurring thematic element in a narrative scheme such as, for example, return journeys, *nostoi*, or heroes. They are widely used by speakers, because they represent a tool in communication that is known and understood by the audience. According to Aristotle's second book of rhetoric, the most effective place to begin an argument are these so-called commonplaces, areas where there is common ground with the audience.

Another important technique is contrast, very effective since ancient times for its psychological force. In the world of wine, there is no shortage of examples: tradition versus innovation, wines aged in barriques compared to those produced in large barrels, native vines vs. international, indigenous and selected yeasts, conventional and organic viticulture. But is emphasizing such contrasts, which appeal to the consumer, especially the less evolved and influential, suitable for providing reliable and transparent information about wine? Does the consumer benefit from communication in which the various aspects of production are placed in a competitive framework or is it better to find a compromise, as often happens in practice? If not, we risk creating divisions and tensions, giving rise to tribes of people who desperately try to defend their positions to the detriment of wine consumption. Instead, we need serious communication, rich in technical content made with simplified language, which leaves the consumer freedom of choice, which is not only emotional. A few guiding words are enough: landscape, territory, history, culture and environment.

A "New World" Between Tradition and Innovation

An impactful narrative model for discussing a wine, region or brand contrasts the past (history) with the present (current reality). Historical reconstruction often draws on familiar symbols, such as classical Greek architecture, mythological references to ancient wines, or ancestral viticulture imagery like bush-trained vines. In contrast, the present focuses on environmental sustainability, global recognition, or innovative practices, such as the cultivation of PIWI vines. This strategy develops the concept of a "new world", where history, myth and contemporaneity come together in a modern wine that targets Gen Z.

The recent presentation of the new Franciacorta Bellavista Alma Assemblage 1 represents an effective example of storytelling, where a meta-story tells how to face and win the challenge to produce a wine of great quality, proposing the image of a "new world", as a promise of balance with all the values of reassurance and peace that identify the soul of the brand (or territory), around symbols that recall universal feelings such as innocence, creativity, harmony, audacity and love. Vittorio Moretti relies on his daughter Francesca, who is helped by a collaborator in the cellar (Richard Geoffroy) and in the vineyard (Marco Simonit), to win the challenge of climate change using innovative (space economy) and traditional cultivation techniques. These include preservative pruning and vinifying grapes that come from many terroirs of Franciacorta, thus enhancing the biodiversity of the territory through sustainable strategies in a subtle balance between tradition and creativity with the aim of giving expression to new trends in taste. To the acquisition of her own skills (degree in oenology at the University of Milan) and of the *chef de cave* of Champagne, Francesca Moretti associates the sparkling wine tradition of the territory, the *stil de vie* of the brand that together identify the main archetypes in the communicative construction of the brand: wisdom, creativity, simplicity and reliability.

The archetype evoked is that of innocence, which emphasizes a desire for purity, good feelings, simplicity, a sense of childhood, something confidently idyllic of the family, that recalls renewal, positive reinvention, entry into a promised land, and transmits a mystical sense of integrity, defined in the synthesis of a subtle balance between tradition and creativity.

How do these evocative elements fit into the storytelling of wine?

The hero is the consumer, and the value is represented by the wine.

The territory is the conduit, the means through which the hero can access the prize.

The Role of Metaphor in Wine Storytelling

A metaphor replaces one term with a figurative equivalent to enhance comprehension. In specialized communication, technical jargon can alienate those unfamiliar with it. Young consumers, for example, may feel excluded by overly complex wine terminology. Metaphors can simplify and clarify the message.

Language must be "common but precise". Metaphor can help. In Plato's *Symposium* (also known as *The Banquet*), metaphors such as curiosity and the half-full glass serve a pedagogical purpose. Common metaphors in wine storytelling include the hero, the border, the journey and anguish. With these metaphors we can tell the story of the many Italian viticultures, the origin of the vines cultivated, and the concern about climate change.

The "Hero" Metaphor

The hero narrative is a canonical storytelling structure, widely used in American communication, especially in advertising for sports or children's products. It involves the hero overcoming obstacles to achieve a final prize. This structure, familiar from childhood stories, comics and films, is central to popular culture, from Harry Potter to Star Wars and James Bond, even Walt Disney's Mickey Mouse is described as a hero of his genre. The American film star Mel Gibson said this about the importance of the hero: "I've always chosen stories about extraordinary characters because that's what movies should do: tell the story of heroes to be inspired by". Even electoral campaigns are focused on candidates who become brands and protagonists of a narrative plot. They must communicate emotions (using feelings such as fear and hope) and values that address the needs of individual voters. In the context of wine, Odysseus serves as the archetypal hero. His journey through the Mediterranean mirrors the voyages of ancient sailors. There are parallels between the stories of Odysseus and that of James Bond: both encounter mysterious women and exotic locations, such as Monte Carlo for Bond or the Cyclades for Odysseus. In both stories, wine plays an important role — Dom Pérignon Champagne for Bond and the wine of Thrace for Odysseus, which he uses to outwit Polyphemus. In this case it is a meta-story that precisely frames the role that Thracian wine plays, as a helper of the hero. The underlying archetype is that of the magic potion which is at the same time a means to increase the hero's strength (like the concoction that the druid prepares for Asterix and Obelix) and is a helper in the fight against the antihero.

A few years ago the sports drink brand Gatorade used Usain

Bolt and Serena Williams as the protagonists of its advertising campaign. The company explained how Gatorade was created and established the Gatorade sport science institute which scientifically studies the effects of the drink on athletes.
An important condition, but also difficult to achieve in the narrative, is the multidisciplinary nature from which to draw the elements to be combined in a story. A story is fascinating if told from a cultural, anthropological, historical, geographical, biological, oenological point of view.
An analysis of the Mueller-Thurgau value chain also offers us the opportunity to identify the themes on which to set the hero's story, contextualizing it in a precise territory. The choice of this wine is not random but paradigmatic because it allows you to bring into play all the elements of the narrative that must interact with each other. The production of Mueller-Thurgau is located in a very specific place, the Val di Cembra in Trentino, and has a strong identity with the alpine characteristics of that territory. It is easily recognizable by its taste-olfactory profile and is a wine for young people that creates an engaging relationship with the target. It also offers a good story. This vine is a cross created by a Swiss researcher Mueller, in Thurgau, a town in German-speaking Switzerland, in the second half of the 19th century, a dramatic time for European viticulture threatened by phylloxera. It has the characteristics necessary to build an intriguing story: the role of genetics, the contribution of molecular biology that has identified the true parents initially misunderstood by the breeder himself, the reference to the alpine environment, pure and uncontaminated, the soils of volcanic origin, the easy recognition and therefore has all the elements to develop the concepts of originality, naturalness, identity and typicality.
In this example, the hero is the researcher who created the vine,

the villian is the public opinion of the time opposed to genetics, the ally is the need to find better performing vines in an era scarred by phylloxera, the prize is the victory of progress over tradition and the favorable reception by German winemakers who made Mueller Thurgau the most cultivated white grape variety in Germany.

Ulysses, the archetypical hero

The prototype hero, Ulysses is the brand of Magna Graecia who maps the way for Euboean sailors. The quest is the difficulties that the hero encounters along the way; the adversary, the hostility of the sea and its deities in which the mysterious adventure takes place. The conflict is the wars between the different peoples of the Mediterranean who fight for the survival of their colonies. The trauma is the violence (the encounter with Polyphemus) to which Ulysses must respond with the strength of the brand and the treasure. The truth event, which helps the hero against hostilities, is the wine of Thrace, symbol of Dionysus. The magical objects are the gifts he receives and brings with him (the wine of the priest Maroni, Calypso's father); the helpers, the women he meets who have contradictory feelings for him (Calypso, Circe). The final wedding is the victory over the prevarication and arrogance of the suitors, the victory of good over evil, of positive values such as friendship, solidarity with companions, affection for the child. In the Homeric poems there are numerous incidents, real *topoi,* in which wine, often sweet, is at the center of the relationship between the characters, representing a unifying element, a sort of magic potion.

Not all heroes embody positive values. The story of Jason and the Argonauts, in which the protagonist leaves the ancient city of Iolcus for Colchis (present-day Georgia) on a quest to find the Golden Fleece, unfolds on two levels: the horizontal one, that corresponds with the trials and tribulations of the Argonauts, and the vertical one, that submerges into the previous heroic mythologies, those of the Homeric saga and the Nostoi. Jason is a sort of "accidental hero", constantly cornered by circumstance, lacking a sense of control, obscured by the semi-divine qualities of his companions. His merits are limited to a certain attractiveness that he is ready to unleash in an unscrupulous way when it comes to saving his skin. In the antithesis of the figure of Odysseus, the myth of Jason finds success and popularity in the normality of his stature: in the end human charm prevails.

Jason, Georgia, the Black Sea, Kvevri, ancestral wine and naturalness occupy a different position to that of Ulysses, the Mediterranean, Thrace, Pithos, amphorae, sweet and aromatic wine. The narration and storytelling take two different paths. One identifies the Greek colonization of the eastern coasts of the Black Sea, the other that of the western Mediterranean. The symbols and archetypes offered by these two hero-travelers, even if they share some common traits (the spread of the myth of wine and the relationship between wine and divinity) are profoundly different and involve universal values such as courage, honesty, generosity, love and innocence. Like the myth of Jason, the recent "return to the amphora" can convey a simple style of unadulterated wine, just like the humanity of the hero.

The biographical drivers on which we build our narratives can successfully use the lives of those who, by virtue of their charisma, have left an indelible mark of their work in thoughts and deeds with prophetic content. The metaphor of Ulysses' journey across the Mediterranean represents a highly effective way to convey the action of a hero. Ulysses is the hero in search of himself, who completes the journey of self-discovery with the help of a treasure or truth event, also thanks to a cast of accomplices. Along the way to reach a goal, the hero must overcome various obstacles, while the final prize to crown the feat is represented by the realization of his dream.

Among modern heroes we can include scientists and researchers. Through their fascinating stories, we can explain the progress made by science as the only remedy against the problems that afflicted our viticulture of about two centuries ago. Today, those challenges are re-emerging, albeit in different forms, with consumer demand for an increasingly sustainable agriculture and the response to climate change.

Mario Incisa della Rocchetta, a modern hero

The Mario Incisa *legend* that has been linked to the narrative of Sassicaia wine since the 1980s, began far from Bolgheri. Its history is intertwined with that of Europe at the turn of the 19th and 20th centuries, where it encountered some of the great calamities and scientific innovations of that period. We should think of Mario Incisa as a modern hero. His is a very topical story, because in his experience as an agronomist and environmentalist, we can identify some effective antidotes to solve the problems that afflict our time, such as environmental sustainability and the consequences of climate change.

It all began in Olgiata, the estate where Incisa moved in the 1930s and where he remained until 1943. Olgiata was his real project, financed by the successes of the great racehorse Ribot. Olgiata was given to Mario Incisa by his mother Eleonora Chigi, belonging to a noble Tuscan family, redeemed from the American funds of Clarice della Gherardesca, wife of Mario Incisa. At Olgiata, a model livestock farm with farmhouses built for employees, an aqueduct, and a dairy, became the headquarters of the breeding of racehorses of the famous Dormello Olgiata breed, including the Thoroughbred Nearco and the undefeated Ribot.

On returning to Bolgheri from Olgiata in 1943, Incisa began the production of fruit and vegetables for Switzerland, and gladiolus bulbs for Europe and America. When Ribot died in 1972, the Marquis's incomes collapsed, and it was only wine that saved his fortunes. And to think, the great racehorse Ribot never made it to Bolgheri!

Not only Sassicaia

From the *brand image* that dominated the market in the 1980s to the *brand story* that has imposed itself since the 2000s, narrative sequences have replaced advertising campaigns. It is no longer enough to make a brand famous or be known to a large number of anonymous consumers, it is now necessary to create a singular, emotional relationship between the brand and its affiliates.

The avenue of cypresses or the beauty of the landscape or the quality of its wines is no longer enough to tell the story of Sassicaia. It is now necessary to give the history of that wine and that territory a new dimension that can excite the consumer, consistent with the feelings that characterize contemporaneity, such as environmental sustainability, respect for nature and its rhythms, the value of biodiversity, which Mario Incisa's work realizes first in Olgiata and then in San Guido, and which he masterfully describes in his book *La terra è viva* ("The earth is alive") published posthumously in 1984.

In this book, Incisa describes those values that we find today in the soul of consumers and the storytelling of his life as an enlightened and curious agronomist proves to be the best vehicle of the "ideology of change", a state of mind that involves the realization that we are constantly on the threshold of change. The essay represents an impassioned cry of condemnation of the harmful effects on plant life caused by deep ploughing that in the 1930s reclaimed the Italian marshes with the introduction of mechanization.

With the arrival of mechanization in the second half of the 19th century and the introduction of the reversible plough capable of "breaking up, turning and smoothing the soil" was introduced, Mario Incisa experienced a profound phase of transformation in the figure of the agronomist and his role in the face of advances in physiology, biochemistry and agricultural chemistry. Certainly, the dramatic effects of the dust storms (the *Dust bowl*) in the Western States of the United States in the 1930s, those of the Great Depression, caused by intensive ploughing, left their mark on Mario Incisa's way of interpreting agriculture and contributed to the birth to a new environmental conscience in him. It was a conceptual revolution, which moved from a "phyllocentric" agriculture (which focuses on the above-ground surface of a plant) to a "rhizocentric" one that tends to focus on the roots with processing and fertilization but through agriculture defined as conservative that aims to avoid the oxidation of organic matter. In 1959, Incisa donated the 90 hectares of the Bolgheri Marshes to the WWF, which together with Burano became the first wildlife haven in Italy. In 1966 he became the co-founder and first president of the Italian WWF.

The Renaissance of Italian wine through the history of the Super Tuscans

The first *Super Tuscans* (the birthright of the term belongs to Frank Pryal, for years wine columnist for the *New York Times*, to give a definition to those unconventional wines), Sassicaia and Tignanello, are the result of the intuition of Mario Incisa della Rocchetta and Nicolò Antinori and the oenological wisdom of Émile Peynaud and Giacomo Tachis. Before them, this innovative way of making a Tuscan wine other than Chianti was shrouded in mystery. Many companies had begun to introduce foreign vines into the Chianti blend, but in an almost clandestine way. At a conference attended by Luigi Veronelli in the early 1980s in San Felice, some companies that exported mainly to the United States began to talk publicly about these wines. Chianti was thrown into crisis.

After the end of sharecropping and with the introduction of the DOC wine legislation, the reconstruction of Tuscan viticulture, supported by the Fondo europeo agricolo di orientamento e di garanzia (FEOGA), started off on the wrong foot. Sangiovese propagation material of modest quality, wide planting layouts, production and vines per hectare too high to provide grapes suitable for a quality wine. Foreign consumers were deserting Chianti. The large companies began to remove the white vines from the famous Ricasoli blend and began to experiment with the contribution of international vines such as Cabernet and Merlot to give Sangiovese greater chromatic stability and an aromatic profile, also with the help of the barrique.

Mario Incisa delle Rocchetta's experience as a winemaker began with the desire to produce a wine to offer on the occasion of the resounding victories of his horse Ribot at racecourses all over the world. He was tired of toasting with the wines of his competitors, especially Bordeaux. He began by planting a vineyard on the eastern side of the castle of Castigliocello, using Tuscan vines, such as Canaiolo, Sangiovese and Trebbiano. The result was modest, the wines were often acescent and Veronelli, his mentor to whom he often turned for an opinion, was very critical of him. A dense correspondence testifies to the sometimes-harsh confrontation.

Chance, or perhaps the meeting with Émile Peynaud who was then a consultant for the Antinori, (Mario was brother-in-law of Niccolò Antinori, Piero's father), convinced him to cultivate vines other than those of Tuscany and a relationship with Count Salviati di Migliarino of Pisa made him aware of the wines of Cabernet grown in those lands. With the cuttings of the vines of the vineyards of the Salviati counts, he planted the first in a plot called Sassicaia, not far from the sea. In 1968, with the help of Tachis and the encouragement of Veronelli, he presented himself at an important competition in London, in which he beat all the best Bordeaux Chateau. The success was repeated in 1985 in a comparison with Cabernets from all over the world.

Where is the novelty? In having won despite the constraint of tradition imposed on Tuscan vines. But above all in communicating without hesitation or sense of inferiority, telling a beautiful story, where together with his audacity, horse racing exploits, innovative cultivation and winemaking experiences, takes on the most important markets in the world. And many anecdotes. A true success story!!

Pierre Viala's viticultural mission in America

Among the modern heroes it is also necessary to include scientists and researchers, to relate, through their fascinating stories, the progress of viticultural science. Among these, Pierre Viala occupies a pre-eminent position. His journey took him to the United States in the second half of the 1800s in search of the wild vines needed to rebuild European viticulture threatened by phylloxera. From 1874, with the planting of the first vineyards on American rootstock, the first serious symptoms of yellowing began to appear, due to the difficulty of the rootstocks then present in France to absorb iron from the soil. In 1887, Viala, a 28-year-old professor of viticulture at the ANA in Montpellier, left for the United States with the sole objective of finding wild species capable of tolerating the limestone present in high quantities in two important French wine-growing regions, Champagne and Charente.

Before starting his journey, Viala stopped for a week in Washington at the Federal Geological Survey to ascertain where the calcareous soils were that might be comparable with the calcareous-marly soils of France. He found something analogous in a territory between Missouri and the Rocky Mountains, from Montana to Texas, whose geological origin is to be found at the beginning of The Mesozoic Era, in the Jurassic. As with Champagne, the soil was formed by the filling of a depression, by the tectonic origin of the Rocky Mountains, by the accumulation of marine deposits and landslides that slipped down the slopes of the mountains.

For six months, from 5 June to 3 December 1887, Viala traveled the United States, but focused his attention on Texas, the last area he visited, where he met the French-Mexican anthropologist Jean-Louis Berlandier, who helped him find those places in central Texas where vines "remain green and vigorous in marly soils". The conditions of the journey were very difficult, involving long stays in hostile and uncivilized places, high temperatures that reached up to 45°C, and encounters with a savage landscape populated by wild animal. The mission report is full of fascinating observations, but above all it managed to bring to Europe numerous seeds of wild vines (*Vitis berlandieri*) that could be used for the creation of hybrids of rootstocks suitable for southern and calcareous environments and which would be decisive for the development of continental viticulture.

It is not difficult to see in Viala's scientific mission in America the narrative model of the hero. In this case it is not a meta-history even if the elements of adventure and risk remain due to the presence of hostile indigenous populations and rattlesnakes. Viala is the hero in search of himself, the anti-hero, in this case, are the opponents of grafting, the obstacles that the hero must overcome (climatic and environmental difficulties), the feat accomplished to discover a treasure or a truth event (the *Vitis berlandieri* tolerant of limestone), also thanks to the help of a companion (the aforementioned Berlandier) and the final prize to crown the feat (the creation of new rootstocks).

The "Border" Metaphor

The metaphor of the border is fundamental to telling how the civilization of wine developed in Europe, and particularly the origins of cultivated vines. Borders are notoriously places of crisis, ethnic tensions, contrasts but also of exchange, regulation of different systems, innovation. It is a cultural paradox that can be at the same time a place of separation (involving the creation of so-called parallel societies) and of encounter and fusion (or convergences).

There are countless elements that define a boundary. In addition to spatial boundaries, represented by obstacles to movement such as seas, mountain ranges, deserts and rivers, cultural, religious and linguistic boundaries are more difficult to overcome. Heidegger in an essay published in 1954, referring to the colonial experience of the Greeks in Italy, wrote: "A boundary is not that at which something stops, but that from which something begins."

To represent something that separates and unites at the same time, it is necessary to identify a sort of no man's land, a place between two spaces, not necessarily geographical, each of which is occupied by a society or culture, distinct from each other. As a rule, the strongest culture has an intrusive action and in its expansion movement presses on neighboring societies and therefore in the case of the vine it is the Roman Empire that exports its viticulture model to neighbouring regions.

In its definition we make no distinction in the meaning between border, frontier and *limes*. A confine is a shared space (*finis*) where two cultures that come into contact integrate to give rise to a new culture. Many grape varieties were born thanks to these interactions. The opposite is the border, where two cultures meet, face each other but do not merge, opposing each other. Over time, the expressions of these viticultural cultures remain unchanged, allowing us to trace their origin. The *limes* is instead a physical limit, where the differences are represented by mountains, rivers and seas. Significant *limes* are the islands or valleys that are difficult to cross and represent "hidden frontiers", places of isolation and conservation. Anthropologists call these cultural expressions "the edge effect".

Italy with its position between West and East, on the path of populations that have always crossed it, with the countless cultural and productive manifestations of its viticulture, is the most effective illustration of the outcome of the encounter between different cultures. Such were the Roman *limes* along the Rhine and the Danube or the case of Termeno in South Tyrol (*Tramin* in German), whose name derives from the Latin word *terminus,* which indicated a geographical and political border for the Romans, in the process of expansion to conquer the rest of Europe. Even the German translation of Collio, "den Ecken", the corner, hides its identity as a place located on the margin, on a border.

Italy is still rich in these borders: just think, for example, of the "hidden borders" that separate Latin viticulture from Greek viticulture on the island of Ischia or from that of the Etruscan enclave of Capua, surrounded by the territories of the Eretriesi. The same can be said of the clear line of demarcation between Lombard-inspired viticulture west of Bologna and Byzantine viticulture up to the Adriatic Sea. Curiously, the border between the viticulture of Slovenia and the Brda does not present significant features of distinction today except for the language, but the sharing of Ribolla as an identity vine of the two territories, cancels the perception of any differences.

In fact, in the search for the elements that allow the identification of the so-called "hidden boundaries", various indicators are used. These include terminology (e.g. the term *karax*, the vine pole of

Massalia, which distinguishes the viticulture of the Greek approach of the "albarello" from that of ancient Liguria, where the plants served as support for the vines); the use of different types of tools for pruning (the sickles and the *pruning knife*) and winemaking, like the castle and the ship for the transport of grapes and wine in the Po Valley; and the types of press.

Also of great cultural interest is the border in the Salento region of Puglia, where different burial habits were practiced on either side of the border. A few kilometers separated two peoples who honored their dead either with incineration or by burying them standing with their faces turned east in pits dug into the limestone. The former symbolically extinguished the pyre with the wine by breaking the *kylix* of the offering, the latter did not include any ceremony. The vines that are still present in the territories occupied by these two peoples are different, as are the winemaking techniques.

Even the dichotomy of the landscape marked by forms of sapling or tree training made it possible to highlight the distinctive features of the viticultural types (vines, the characteristics of the wines) of the peoples who had come into contact.

The island of Ischia is exemplary in this regard. On the two geological matrices that characterize it, the tectonic one, the result of the emergence from the sea, the Greeks who settled used the clays of flysch to make amphorae, while on the volcanic one, an expression of the intense volcanism of Campania, the Etruscans / Romans worked the iron that came from Elba. From this side, from the Neo-Latin linguistic connotations, at the end of the historical cycle of the Bronze Age all the ships that brought iron to Greece for the manufacture of weapons departed. Where the Etruscans worked, the vine was grown on the phalange, very high chestnut supports, in the fashion of paleo-Ligurian viticulture, while in the part of the island inhabited by the Greeks, where there are also lexical sub-

strates of the eastern Mediterranean and the Caucasus, prior to the Greek one, the vines were pruned to sapling or low espalier, locally called "latina". These two different forms of training were divided by a road, which still exists today, which has made it possible to keep the original substrates miraculously intact.
From a methodological point of view, an ancient grape variety, even if no longer cultivated and present as a relic in a collection, can represent an effective indicator of the existence of a cultural boundary in European viticulture.

Borders and the origins of cultivated vines

In countries with high immigration, such as the United States, the phenomenon of searching for one's family roots is growing strongly. Usually this interest appears in people, around the age of sixty, when they begin to glimpse the end of existence and in times of wavering faith, they feel the need to reconnect with their lineage, to reaffirm that we are something more than an individual. In the past, research into family histories required long and arduous journeys. More recently, the volume of information available in the registries of municipalities and parish registers, as well as in the DNA databases of relatives, has increased. A visit to the places from which the ancestors left remains essential to grasp important details and discover the territories where there are "invisible" roots, which time cannot erase.

For the vine, the first years of the new millennium represent the watershed between the ampelographic knowledge of the 19th century, lacking objective tools for the recognition of vines, and the results of DNA analysis that will inaugurate an unexpected phase of discoveries that will tear through that veil of mystery that until then, surrounded the origin of vines. With the irruption of molecular genetics in the recognition of vine varieties, many myths and fairy tales born in the idealist period on the origin of vines have been revised. Above all, for many varieties, it has been possible to trace their geographical origin, reconstruct the paths that led them to us, identify their relatives, even distant ones, who contributed to their formation, highlighting the decisive role of culture and the original aspects of the material life of the various populations, in their selection, protection and subsequent dissemination. Through the behavior of these vines towards the environmental conditions and the wines that give rise, it is possible to understand the climatic phase in which they were selected, the extent of winemaking interventions, and the expectations of consumers of that historical period. There was no shortage of surprises. Close to unsuspected parents and fathers who are not easy to identify, what appears increasingly evident is a highly varied network of relatives where a very few vines represent the progenitors around which most of the European vines are formed.

The traces to be followed to reach the truth, with apparently dichotomous meanings, originate from signs that must be deciphered with the help of anthropology and molecular genetics, very different tools of investigation but curiously complementary in their results.

For European viticulture, the boundaries between peoples and cultures, often marked by strong contrasts and disputes, have been privileged places for the birth of new expressions in the cultivation of vines. From peripheries they have become centers of gravity. The Greeks, first of all, with both colonial and commercial experiences (*emporion*) contributed to the spread of the myth of wine in the West. The Romans, with their military conquests, brought viticulture to the border regions of the empire, where the demand of the legions for wine was greater, resulting in the birth of new models of viticulture.

Due to a series of climatic and commercial circumstances, most of the vines still cultivated in Europe originated on the eastern borders. It was a slow, reiterative process, with which the DNA of some vines brought from the eastern regions, where viticultural practices, such as pruning and genetic selection, were introduced into the vines derived from the domestication of the wild vine, which took place starting from the Neolithic on the banks of the great rivers. An important aspect in the creation of new vines was that of the ways in which the Romans involved the local populations, not as slaves but as allies, through the authorization that the Senate had granted to cultivate vines in their territories, a prerogative then reserved only for Roman citizens.

The role played by wild vines domesticated on the banks of the Rhine is significant. Traminer, for example, which through spontaneous crossbreeding with an oriental variety, Heunisch, contributed to the creation of numerous vines that are still important today for the quality of their wines, such as Pinot, Sauvignon, Riesling etc., and through these with successive cycles of crossbreeding Cabernet Sauvignon, Merlot, Carmenere etc.

The contribution of anthropology in the narration of the origin of vines

The study on the origin and diffusion of cultivated vines represents the interpretative paradigm of the findings of recent studies on an important topic of vine history. To demonstrate how anthropology, with its methods, appears decisive in providing the solution of otherwise indecipherable questions, even if addressed with the most advanced tools of viticultural research.
If we start from linguistics research, the line of reasoning seems more linear. Today, more than five thousand languages are spoken in the world, all of which are the result of three fundamental processes. The first is colonization, as a result of which the ancients, penetrating a new area, introduced their own language. The second is divergence: the language spoken by individual population groups changes over time. The third is substitution: a group of people adopts a new language, introduced or imposed by another human group. These three processes make it possible to draw suggestive schemes in which the origin of the languages currently present in the world can be arranged.
Varietal circulation has also followed these patterns of diffusion. Despite the variety of approaches, a constant trait emerges from research on the identity of the vine in different parts of Europe. The origin of a continent without rigid borders that has always been a place of migrations, interactions, hybridizations, contrasts and conflicts between peoples and that has drawn essential nourishment from the diversity of roots on a cultural and political level. The study of kinship relationships between the vines grown in Italy has highlighted an important genetic link between the populations present, to such an extent that we can reconstruct the pedigrees of the vines present in Italy on eight vines, defined as founders. The

close genetic connections between the vines of Greek origin and those present today in the regions that in the past were ascribed to Magna Graecia, however, do not only date back to the colony or commercial experiences of the classical period, but also originated from the subsequent cultural contribution, Byzantine and Basilian. The political and social situation of the 5th and 6th century A.D. contributed to a profound demographic crisis in southern Italy, which was faced, especially on the Ionian coast and the lower Adriatic, with strong flows of emigration from the Middle East, Syria in particular, and the Balkans. Byzantine influence became increasing strong, especially at the level of secular ecclesiastical structures, with a process of progressive Hellenization in those areas where hermits and monks of Eastern origin had settled. These communities deserve credit for having tilled many uncultivated lands and for having removed the local populations from the "absolute cultural primitiveness", for having educated them in agriculture and in the artistic taste of the East, as well as of course in the language. Above all, the monks of San Basilio are credited with having organized these populations into agricultural communities with a strong religious element, introducing new species of animals from their places of origin, including silkworms and arboreal and herbaceous plants such as mulberry and sugar cane.

Calabria, in particular, was affected by the Byzantine migration phenomenon, as evidenced by the now modest traces of the luxuriant vineyards of Bivongi, Stilo, Brancaleone and Rossano perched on the steep slopes of the hills, where they could defend themselves from the raids of the Arabs. Latifundium and malaria of the coasts had already depopulated the plains. This viticulture was, therefore, no longer only of Magna Graecia origin, but also Byzantine, and the many varieties that still survive, although as relics, are the testimony of that period and the places where the Basilian monks came from. Certainly, some of those varieties had been transported from the regions of present-day Bulgaria, Romania, Serbia and Croatia and are therefore different from the vines of eastern Greece and the Black Sea coasts, such as the white Mantonico.

The biological diversity of the cultivated vine, the result of thousands of years of selection, is a legacy that nature and our ancestors have left us and that cannot be recreated in a laboratory. Once this resource is destroyed, it cannot be replenished and will be lost forever. But this diversity not only has a biological value as a phase of a natural, albeit human-driven, evolutionary process, but it is also an economic resource that arouses the growing interest of the consumer in wines produced from these "minor" vines, as well as the many economic activities, from tourism to food, associated with it. The complex origin of the vines can represent a symptomatic element of a way of thinking about a given area. The vine can be placed at a symbolic level without placing importance on the production process. It is important, therefore, to understand why this process took place in this or that region and what significance can be attributed to it.

Myth, according to the principles of Claude Lévi-Strauss's structural anthropology, becomes the methodological tool for conceptualizing the problem, as it allows the history of a group of inhabitants to be transferred to the present time, in a tangible way. For a given type of wine, the myth of origin or domestication makes it possible to explain the genesis, evolution and originality of production. The arrival of the oriental populations in Europe during the countless invasions is configured as a real revolution that has iterative characteristics that involve a slow but continuous change, just like the birth of a vine.

Even the word "autochthonous" had gradually lost its meaning. Vines are the result of an intense and ancient varietal circulation even between very distant areas, often without geographical boundaries. The term autochthonous, therefore, no longer relates to a place, but to the time in which a vine manifests itself in an optimal way through its production characteristics. With the systematic application of nuclear DNA analysis through microsatellite markers, important links have been established with the viticultural germplasm of some regions. For example, Veneto, Trentino and Friuli, through a genetic relationship of Corvina and Marzemino respectively, with Refosco dal peduncolo rosso or the links still present between the DNA of Calabrian and Sicilian vines with that of Greek or Anatolian varieties.

Culture and the boundary between wild and cultivated vines

In July 1774, the German botanist Samuel Gottlieb Gmelin was kidnapped in a remote part of the Caucasian mountains and subsequently died in captivity. Despite his young age (he was barely 30 years old when he died), he had already undertaken extensive scientific expeditions in the regions around the Caspian Sea and was responsible for the placement of the wild vine in the context of the Linnaean systematic classification of *Vitis vinifera*. The student who assisted him in his final hours reported that in his agony, the German scholar recited the Latin names of the new species he had identified, including the wild vine.

To tell the truth, since the first Greek geoponics, the distinction between spontaneous and cultivated forms of vines has always been clear, but the relationship between the two subspecies has always been ambiguous. Often the only element that allowed a correct taxonomic attribution was represented by the different era of domestication that profoundly affected the phenotype of the vines that had been absent from the condition of wild plants for a long time. This observation rejects the categorization of cultivated and wild vines into two distinct subspecies and considers them two chronological stages of the same evolutionary path.

Developments in molecular biology and hypotheses on the origin of human populations have recently confirmed that the varieties cultivated in the West are not only the result of migration phenomena but also the result of punctiform phenomena of para-domestication that have in very distant times, generation after generation, acted on wild populations by shaping their genetic structure.

The metamorphosis of the vine from a wild plant to a cultivated one is the exclusive result of mankind's demiurgic activity. In the

slow progress of genetic selection and in the improvement of its qualitative characteristics, humanity has mirrored itself, domesticated itself, and established a paternal relationship with the vine.
In the face of a natural genetic variability of spontaneous populations, not always very high, the presence of thousands of vines in European viticulture, testifies, close to the important role of gene introgression, to the essential contribution played by the peoples who domesticated them, guided by their culture and their myths, starting from the choices of their daily life.
Paraphrasing a statement by Louis Levadoux (1956) the secret of the origins of our viticulture is in the wild vines. In the analysis of the individuals still present in the places of primitive domestication, still frequent in central Italy, comparing them with the ancient vines of those areas, it is perhaps possible to reconstruct their geographical origin, the phases of their cultivation, and their subsequent migration to other places.
The beginning, at least literary, of the viticultural history of the West Mediterranean in general, is contained in Canto IX of the *Odyssey*. From the description of the island of the Cyclops and the drunkenness of Polyphemus, it is apparent that since then there have been two viticultures in the Mediterranean. The first, located in its eastern part, which produced "strong and dark" wines, which was specifically the wine of Ismaro, given to Ulysses by Maron, priest of Apollo. It was this wine that made Polyphemus drunk because he was not used to such strong wines. The second, represented by the island of the Cyclops, "who plant nothing with their hands nor plough; but all these things spring up for them without sowing or ploughing, wheat, and barley, and vines, which bear the rich clusters of wine, and the rain of Zeus gives them increase" (Odyssey, IX, 108-11), and which clearly configures the characteristics of wild grapes, subjected to primitive or no cultivation.

Homer's description, however, is not only an important document, the first in chronological terms to understand how the myth of wine spread in the western Mediterranean, the role played by the sagas of the founding heroes and the return voyages (the *nostoi*), but it also takes on the value of a metaphor, that of the coexistence of two models of viticulture that have survived until now in many areas of Italy. They are expressions of very different cultures which have used very distinct vines and forms of training to obtain wines that are still easily distinguishable today.
In the wake of Ulysses' ship, other Mycenaean and Phoenician navigators and the first Greek settlers spread pruning and vine training techniques known only in the Caucasus and the Levant with oriental vines. One can imagine that the Homeric text reflects the production of a wine, in the places mentioned in the epos of the Odyssey, on the Tyrrhenian coast (Gulf of Naples), the Aeolian Islands and north-eastern Sicily. In fact, a part of Lucania and Puglia was already called "land of wine" or rather "land of vine poles" (*enotria*) and this shows that the production of wine in these areas was prior to the foundation of the cities that were protagonists of the *nòstoi*. The vine, in fact, was present in southern Italy before the arrival of the Greeks, in so-called anthropophilic forms of cultivation, especially in the places of Villanovan civilization and Etruscan expansion and most likely was the result of the domestication of wild vines, spontaneous in the lowland woods or the result of sporadic contacts with the populations of the eastern coast of the Adriatic and Ionian Seas.
The introduction of new vines in Roman Italy was a very frequent phenomenon due to the intense political and economic contacts it had with the oriental peoples, with the aim of imitating their wines, which were highly appreciated especially in the imperial era. New varieties spread with ease because of the principles that in-

spired the work of the Latin Georgics, for whom the quality of the wine was not the expression of the genetic qualities of the vine, but of the environmental characteristics.

Among the events that favored the arrival of vines from other places, there was also the solicitation that Emperor Marcus Aurelius Probus had, from his wife of Aquitaine origin, Faustina, who, on behalf of her fellow countrymen, asked for the annulment of the edict issued about 200 years earlier by Emperor Domitian which limited the cultivation of vines outside the borders of Italy. As chance would have it, the legions most loyal to Probus, who were asked to rebuild the viticulture of the *limes*, were located in Pannonia, the soft underbelly of the defenses against invasions, and in this region there were highly productive vines resistant to cold.

Many nameless varieties were brought along the banks of the Rhine and the Danube to be cultivated together with spontaneous vines, which had long been subjected to proto-domestication processes and which had initiated the local populations to wine, although of modest quality. These varieties were generically called "heunisch" (or "hunnisch", from Huns), "Gouais" (in the French spelling with a derogatory meaning), because they came from the territories that were inhabited by the Huns. Chardonnay has its distant origin in this event.

There was no planning in Probus's intervention. He acted without realizing what he was making his legions do. Most likely if he had gone to take the vines to be brought along the Rhine to other regions of the vast Roman Empire or if his wife instead of being an Aquitaine had been a Dacian, the course of the viticultural history of Europe would have taken another direction and the wines that today cheer our tables would certainly be very different.

The storytelling of autochthony: why a new narrative is needed

The term glocalization, coined by Roland Robertson in 1992, defines the ambivalent articulation between the global and the local, between the interdependence of the world-economy on the one hand and the particular dimension of local contexts, with their traditions, on the other. These two realities are both contrasting and complementary, and connote not only a set of economic-financial activities, but also a complex of symbolic, practical, social codes and cultural habits linked to territorial communities. Agri-food production and wine production, in particular, represent the interpretative paradigm par excellence of a need for identity and recognition that arises from the fear of anonymous social ties in an increasingly globalized world and that finds in tradition, no matter if reinvented or remythologized, the anchorage of affectivity or belonging, as the French historian Jacques Le Goff defines them.

The return to native vines and typical products of Italian food are the evidence of this attitude, but it is necessary to remove some misconceptions. Italy is the second largest producer of tomatoes in the world, after the United States. The tomato and its derivatives are a fundamental ingredient of Italian cuisine and a symbol of Made in Italy, so ingrained with our food traditions that we consider them a native product of our country. Yet, tomatoes came to us after the discovery of America. A similar consideration can be made for the vines that are grown in Italy. Most of them came from other countries, often giving rise to new varieties through spontaneous crosses and only a small number are the result of the domestication of the wild vines of a specific place.

Varietal circulation in Italy is a very ancient phenomenon shaped by its barycentric position in the Mediterranean between East and

West. The first evidence of the arrival of "foreign" vines dates back to the great wave of migration of the Caucasian populations that took place during the Early Bronze Age from around 3,000-4,000 BC and may have been caused by the disastrous impact of the so-called Black Sea deluge. The greatest varietal contribution, however, took place during the first millennium BC, with the Greek colonization in southern Italy. The use of autochthonous as an adjective to describe vine varieties is quite recent and is the result of a cultural attitude, called indigenism. This current of anthropology and archaeology aims to treat each society as autonomous and therefore to see its developments in essentially endogenous terms.

We now define as "Italian" many food products that have become "Italian" in relatively recent times or those that have undergone profound structural and sensorial changes with the transformation of modern processing techniques. Not to mention the fruits and vegetables that we consume today that perhaps only have the word Italian on the label. Returning to wine and, in particular, to what is produced in our south, it is noticeable that modern techniques to make it fragrant with flowers and fruit or fresh for an acidity that it does not possess naturally or worse to give a richer and more concentrated structure, make them lose that identity that makes them recognizable and inimitable.

The existence of an important genetic variability in Italian viticulture, although economically negligible, finds its roots in peasant traditionalism, an obligatory choice of rural communities in the past in order to deal with the risk associated with the unpredictability of environmental conditions, typical of the farmer's profession, with inadequate tools. Where people live on the margins of subsistence, there is no place for experimentation, especially genetic experimentation: a risk that can have catastrophic consequences. These simple considerations show how the evolution of ampelo-

graphic platforms has only recently been the result of the intertwining of economic and cultural behaviors, while major genetic modifications have occurred mainly for causes beyond the control of winegrowers and linked to major climatic changes.
Adjectives used to describe grape varieties, such as minor, local, ancient, autochthonous, old, domestic or indigenous, are often used incorrectly to define what may seem like the same thing. In fact, a local grape variety is not always autochthonous just as a minor grape variety is not necessarily ancient. The return of ancient or native vines to the global stage has brought back into vogue the crucial role of the grape variety in the communication of wine in an era in which there is no time for geographical insights and the role of the aura that accompanies its origin and history, compared to a consumption made more with the brain than with the palate. Of these vines we often remember only their name because of curious vernacular. We have already forgotten, after a few years, the taste sensations of their wine often produced with oenological techniques that are not consistent with the characteristics of the vine and the remote location where they are grown.
What should we do, then? What path to take in a period of great changes in terms of consumption, lifestyles and last but not least the climate?
The pervasive element in every choice, whether it is productive or consumer, is globalization. Accepted by many for its positive contribution to cultural complexity and commercial relationships, by others it is considered disturbing and dangerous and therefore rejected. It is widely believed among producers and consumers that it is the main culprit of the increasing simplification of the sensorial characteristics of wines produced in all parts of the world, strongly conditioned by "perfect" types of wines, but lacking in soul and unable to recognize "praise of imperfection". In this sense, the role

of communication and guidance is also important. Imperfection is not a defect but often the basis of a wine that goes against the tide, of an innovative wine.

But globalization has also operated in the opposite direction, creating new opportunities for alternative forms of viticulture, such as biodynamics, in an attempt to maintain expressions of wine production that the pressures of scientific progress tend to upset. But turning to "new age" philosophies to produce wine does not make much sense, nor does taking refuge in a tradition that is seen as the antidote to the evils of research. Nothing could be further from the truth, because as Hobsbawm and Ranger said in *The Invention of Tradition*, traditions have always been invented and reinvented to satisfy the purposes of people who wanted to use them to legitimize their power.

As a general rule, a local variety is also ancient because it has been cultivated in that place for a long time, although it seems problematic to specify the time scale due to the uncertainty in defining the time of introduction. A variety can be very old but not local. This is the case of vines from the East, which have long been cultivated in places far from their places of origin, such as Moscato and Malvasia.

Often the difficulties encountered in tracing the origin of the varieties are also due to the change that the name of the vine undergoes as it passes from country to country and undergoes a sort of vernacularization that makes it local through the use of dialectal expressions. Local varieties often take on names that derive from specific characteristics of their morphology or productivity (vernacular name) such as the color of the berry, the habit of the vegetation, the shape of the bunch, the ripening period or the place of origin. This is because the selection of these vines, whether of local origin or introduced, was strongly integrated with the traditional

agricultural system and made the subject of particular cultivation practices. Moreover, the wide intravarietal genetic variability, if it had allowed them to adapt better, would have limited their interest if they were separated from their environmental and anthropological context as an expression of cultural peculiarities of the agricultural society (food aptitudes, local needs) to which they belonged.
If autochthonous is a concept linked more to space (as that vine is native to that place), than to time (how long that vine has been cultivated in that place), one wonders how long it takes for a vine to be considered an expression of a place and therefore in all respects autochthonous.
In recent years, on the wave of renewed interest in such vines, there has been much discussion about the definition of autochthonous. It is therefore legitimate to question what "autochthonous" actually means in relation to a grape variety.
If we stick to its etymological meaning of "a vine born in that place" perhaps in Italy we do not have more than a dozen native vines. In fact, most of them either came from other countries or are the result of spontaneous interbreeding phenomena. It is probably not possible to give an unambiguous definition of autochthonous viticulture for Western European viticulture, except perhaps for those vines derived from the domestication of local wild vines (such as, for example, Lambrusco di Sorbara, Lambrusco with jagged leaves, or Enanzio, Refosco, Asprinio and Savagnin) or for the varieties obtained from them by genetic introgression (the vine of ancient Pannonia, Gouais in the creation of the vines of continental Europe). Much better would be to replace the word autochthonous with that of ancient, local and Italic.
Is it necessary to know the geographical origin of a grape variety to define it as autochthonous?
Sangiovese is the progeny of Strinto porcino, an ancient vine of the

Vesuvius area and has family relations with Visparola and Ciliegiolo, its son. It has shared kinship relationships with the cultivars of four Italian regions, Basilicata (with Stampacavallo and Strinto porcino), Calabria (with Castiglione di Bova, Gaglioppo, Toccarino), Puglia (on Negrodolce) and above all with Sicily (Nerello, Arbanello, Frappato, Lievuso, Lucignola, Orisi, Perricone). In which Italian region can it be considered autochthonous?
The true meaning of the word then is not linked to the place of origin but where the vine best expresses its genetic potential. Sangiovese is therefore native to Tuscany, to Montalcino, to Chianti, to Maremma and to Montepulciano.
Gone is the era of stateless vines, also known as international varieties, now cultivated everywhere, whose wines are increasingly destined for the lower end of the market. Gone is the confusion generated by a plethora of designations of origin without traditions or qualitative merits. What is the future, then, for Italian viticulture?
In determining the quality of a wine, the still unresolved quarrel over the primacy of the grape variety over the environment and vice versa now belongs to the history of viticulture. Until the "Age of Enlightenment", the place of production prevailed over that of the vines. Jules Guyot was the first in 1868, on the threshold of the phylloxera crisis, to break this axiom with the revolutionary observation that "the genius of wine is in the grape variety".
To those who invoke the return of tradition in wine production, however, we can answer that the most effective way to honor it is that of a "faithful betrayal". The interpretative paradigm that has driven viticultural progress over the centuries has been genetic innovation, represented by varietal circulation and the selection of intentional and spontaneous crosses. After about two decades of great interest in native vines, on the part of wineries, research and consumers, taking stock of the results achieved in terms of eco-

nomic repercussions, it must unfortunately be noted that in terms of enhancement of the vines and territories concerned, these do not appear particularly exciting. There are, of course, some exceptions, which see a few vines achieve good results, but unfortunately many others face oblivion. It is necessary, therefore, to disaggregate the factors so as to better direct efforts in the future.

In summary, of the thousand and more native Italian vines, the quality ones (those used to produce a wine intended for a modern consumer) are very few, perhaps no more than a hundred. This does not detract from the fact that all of them have great value for the conservation of biodiversity and for use in genetic improvement programs. Only a few grape varieties, supported by successful DOC or DOCG, such as Nebbiolo for Barolo, Barbaresco, Valtellina or Sangiovese for Brunello, Chianti, Nobile, to name a few, are expanding. Some ancient vines have been enhanced, not thanks to the institutions but to the passion of a few winemakers (such as Timorasso by Walter Massa, Teroldego by Elisabetta Foradori, Sagrantino by Marco Caprai). Few vines have emerged from oblivion in recent years to be registered in the National Register of vines authorized for cultivation, the sine qua non condition for being multiplied and planted.

“International” grape varieties and the story of modern viticulture in Italy

The presence of the so-called international grape varieties is so important in world viticulture that consumers do not consider them a particular category, because they have penetrated so deeply into the wine culture that they shape the images and sensations we have in our heads of a Chardonnay or a Cabernet Sauvignon. The definition of international grape varieties in the language of viticulture, not only Italian, is quite recent and dates back to the 1980s when a great project was launched in Italy to recover and enhance the hundreds of native varieties present but practically unknown throughout the country. It is from the almost competitive contrast of these two great varietal types, that their compositional differences emerge, the different contribution to the organoleptic characteristics of the wines, to a presence that had previously been almost ignored.

Varietal circulation is a very ancient phenomenon in European viticulture and, if we exclude the contribution of ancient Greece, we can say that this silent invasion began during the Middle Ages with the introduction by the Venetians and Genoese of some varieties suitable for the production of sweet wines (Malvasia, Moscati, Vernacce), and continued later with the appearances of some Catalan and Spanish varieties (Grenache-Cannonau, perhaps Vermentino) in some central-southern regions. At the beginning of the 1700s there was the first interest in French wines, in Tuscany for the production of “*claretto* alla Franzese” and in Friuli with the introduction of Pinot Noir. However, it was the Medici in 1773 who imported large numbers of vines from different European countries to evaluate them on their estates. However, the judgment on the real value of these varieties is critical.

It was in the 19th century, however, that the presence of foreign vines which came from other countries, France in the first place, became important from the point of view of their role in the quality of Italian wines. In 1903 Mondini published *I vitigni stranieri da vino coltivati in Italia* ("Foreign wine vines grown in Italy"), a complete review of the presence in the various regions of the peninsula of the vines that during the 19th century began to play an economic role and were not only present in ampelographic collections. The main reason for these introductions, first made by some important wine producers of central-northern Italy, was to emulate the success of French wines in the world, then the most important economic-military power in Europe. The goal was to feed a current of export of bottled wines to replace the trade of blended wines that were highly appreciated on foreign markets, but not very competitive from an economic point of view. In the 1800s, France became the leading country in the production of wine and in the development of new viticultural techniques which began to permeate the Italian production world around after the second half of the 19th century with the unification of Italy, both for the role that Napoleon III's France had played in the Wars of Independence and for the looming need to renew viticulture after the ailments caused phylloxera.

The first innovations transferred from France a few decades before the arrival of the so-called "American diseases" were represented by certain vines, especially of Bordeaux origin, which were imported by some Tuscan and Venetian landowners to imitate those "French" wines, the clarets that were so successful in England. The first ampelographic collections, expressions of nineteenth-century collecting and the birth of the schools of oenology, contributed to the knowledge of the behavior of these vines in Italian environments and to the correct vinification of their grapes.

The first Cabernet Sauvignon plantings date back to 1820 and were made near the town of Alessandria in the region of Piedmont. The most important noble families from Piedmont, Tuscany and the Kingdom of the Two Sicilies, often related to the French court, who consumed exclusively French wines, began producing these wines from the end of the 18th century. In Italy there was never any talk of wines destined for foreign markets. At this time, the newly founded Accademia dei Georgofili stimulated, as part of a competition, the publication of works of wine erudition to improve the quality of Tuscan wines in order to promote their export. The winner, Dr. Cosimo Villifranchi, bitterly denounced that in Tuscany, and elsewhere, the quality of the wines, whose production was inspired by the French model, had no link to the famous wines of Italic antiquity, thanks mainly to inadequate winemaking techniques ("in Chianti noble grapes, plebeian wines"). In the instructions he provided for producing the best quality Chianti wine, however, he also fell into the tradition of long macerations with the pomace responsible for the acescence of the wines.
In 1825, the Giuseppe Acerbi, a writer and naturalist from Mantova, considered the first Italian ampelographer, compiled a catalogue listing as many as 903 foreign grape varieties in his collection, mostly from Austro-Hungary, of which Lombardy was then part. In 1852, Marchesi Incisa created an important collection in Rocchetta Tanaro, but he did not limit himself to simply collecting foreign vines but also evaluated their viticultural and oenological value. Cabernets, Verdot, Gamay, Pinots stood out for their quality. These collections were the first of their kind in Italy, arousing in many winegrowers a curiosity and desire to experiment.
What motivated such winemakers to introduce these new varieties? The desire to improve the quality of their wines, which native vines often did not guarantee, especially compared with other countries

(especially France). The possibility of extending the cultivation of vines to less favored places where native vines did not yield good results. The hope of being able to cultivate vines more tolerant to cryptogamic diseases, (downy mildew and powdery mildew had recently appeared) and adverse weather conditions.

The pace of change accelerated sharply in the second half of the 19th century with the arrival of phylloxera, which required the reconstruction of Italian viticulture on American rootstock, presenting the opportunity to spread those foreign vines that had given good results in previous years. To dispel the complexities, Dr. Jules Guyot, the French physician and agronomist, made a declaration, revolutionary for the time, proclaiming that "the genius of wine was in the grape variety", thus belying two thousand years of Georgic culture, which saw terroir as the determining factor behind the quality of a wine.

The dramatic relationship that was established between European viticulture and American diseases on the threshold of the twentieth century was the basis of the birth of the so-called modern viticulture. The places of viticulture changed, the specialized vineyards, ordered in flourishing rows replaced the wooded fields, where mulberry or maple plants with married vines alternated with fields cultivated with meadows or cereals, especially the vines. Those varieties that had only the characteristic of producing a lot of grapes and being particularly rustic towards the climate were definitively abandoned.

On the threshold of post-phylloxera renewal, however, the problem of the backwardness of winemaking techniques, which were not suitable for enhancing the qualitative characteristics of French vines such as Chardonnay or Merlot, manifested itself in all its gravity. The cultivation of foreign vines in Italy on the threshold of 20th century remained a sporadic and pioneering phenomenon carried out by some more advanced entrepreneurs, in Trentino Alto Adige (or Southern Tyrol as it was then called) and in Friuli Venezia Giulia, due to the presence of some research and training centers. The replacement of the old varieties with vines of French and partly German origin, favored by post-phylloxera reconstruction, became a generalized phenomenon that led in a few years to the replacement of dozens of old native vines. Their diffusion was initially highly sporadic due to their low productivity, much lower than that of "homegrown" vines and to enhance the quality of these vines, though winemaking operations were increasingly modernized.

This marked the transition from a production of wine mainly intended for self-consumption to that for the market. The cultivation of "French grapes" in Italy had as its objective the production of "French claret" wines, in imitation of the Bordeaux clarets that were so successful on the English market. Very often, however, the vine was used as an improver of a local variety lacking color and aroma. Cabernet Sauvignon was the most represented variety. In Piedmont, the Marchese Leopoldo Incisa della Rocchetta introduced this vine in 1860 in his ampelographic collection of Rocchetta Tanaro and wrote verbatim "... it is a vine of ordinary robustness, but vigorous, which loves a more fertile soil than Cabernet Franc and gives a uniform production every year. The grape has a distinct, but somewhat tart flavor. The vine is to be considered among the most valuable and worthy of being propagated."

Between reality and imaginary: the metaphor of the wine-growing landscape

The great history of Italian viticulture is ultimately the result of many small stories, yet to be told, of an archipelago of communities, resonances, cultural islands and distant villages that were formed by the stratification of experiences and peoples around places of great transit and trade. To these populations, who have preserved intact the immutable rites of pruning, harvesting and winemaking, we owe the oldest winemaking traditions in Italy. The wine-growing landscape will increasingly become the essential vector of knowledge of the vineyards and wines of an area and therefore the most important resource for all oenological and cultural strategies.

The metaphorical potential that a vineyard possesses is very strong. This transfer of sensations from the concrete landscape to the imaginary is a common procedure operated by a vineyard. The vineyard is first and foremost a metaphor of great balance: for the image that emerges from a man-made nature, of a harmony between man and plant, a sort of complicity. But it is also an eloquent metaphor for dynamism. The landscape is a bearer of enthusiasm, but at the same time of rigor and stability that comforts and stimulates the consumer. Beware, however, of the negative implications of a poorly respected landscape or a poorly maintained vineyard: they can express an unresolved conflict between modernity and nature or be bearers of a negative message.

However, the natural elements of the landscape acquire meaning only in relation to historical and cultural contexts. So, the landscape is a constructed space, where the work of man becomes part of it. The landscape is not just what you see. Territorial phenomena are governed by latent structures that do not appear to the eye and must be investigated with appropriate tools.

The landscape must testify to the authenticity of a wine. The word "authentic" comes from the Greek, *authentikos*, acting on one's own authority (*autos* = oneself and *hentes* = being). Authentic therefore means essence, "true interiority", not appearance. Therefore, in relation to wine, authentic means genuine, the expression of a well-defined, defended territory.

Abstracting ourselves for a moment from the image of nature and the landscape in its evolution according to Western aesthetic taste, we can identify the vineyard landscape around an inhabited nucleus, from an eminently anthropological point of view, as a eutopian garden (good place), resorting to myth, to found on this, the memory and identity of its inhabitants.

In the shield that Hephaestus adorns with scenes of country life for Achilles, we find the description of a garden-vineyard: "Therein he set also a vineyard heavily laden with clusters, a vineyard fair and wrought of gold; black were the grapes, and the vines were set up throughout on silver poles. And around it he drave a trench of cyanus, and about that a fence of tin; and one single path led thereto, whereby the vintagers went and came, whensoever they gathered the vintage." (*Illiad* XVIII, 561-566). How can we not see in this Homeric description of the garden of the Greeks, all the characteristics of the Italian "vineyard garden" (aloe, the field planted with vines, *kepos*, the tree-lined and orderly garden, the *horto*, the luxuriant vineyard, surrounded by a wall or a hedge)?

The training system, a hidden border of Italian viticulture

In the collective imagination, the word vineyard evokes a plot of land intensely cultivated with vines, with orderly rows, with vines supported by poles and iron wires. If we go back in the landscape memory of a few decades ago, before the 1960s, as a result of the "great transformation" that sees the disappearance of sharecropping in many areas of central Italy, we realize that the concept of vineyard has changed considerably, as the dominant landscape was then represented by promiscuous crops, in which the vine was "wed" to the tree and the olive tree, and was associated with herbaceous crops.

Few words relating to wine have the evocative power of *enotria*. In the past, this Greek word, which was erroneously translated as "land of wine", generically identified Italy. Instead, "*oinotron*", means "vine pole", and was the name with which the first Greeks who landed in Basilicata in the eighth century BC called those places, defining as *oenotrian* the populations who had peacefully welcomed them.

Why did the sapling with the pole strike the curiosity of the Greeks to the point of becoming the defining feature of those places and that people? The explanation is very simple and is, even today, under the eyes of those who go to the Greek islands, such as Santorini or the Turkish hinterland of Izmir, where the vine is cultivated as it was 3,000 years ago with creeping forms of cultivation.

Low-stock forms were not the only ones present in the most ancient areas of the Greek colonization of Italy, as demonstrated by numerous literary sources that describe in classical times, in Campania, vines grown on live stakes in Campania, similarly to what was happening in a more widespread way in central and northern

Italy. Also significant is the dichotomy in the landscape represented by the Aversa trees, an expression of Etruscan viticulture, and not distant from the alberello, a sign of the cultivation of the vine according to Greek models.
From the 8th century BC, the Etruscans of Capua and the Chalcidians of Cumae established a "hidden frontier" between the two cultures: that of the vine trained as a sapling and that on a live stake. In the countryside of Capua, in fact, from the first Etruscan expansion to the present day, the vine was wed to the poplar and, with its decisive imprint on the agricultural landscape, created a profound dichotomy with respect to the vine grown at low stumps in Campania south of Naples.
Greco di Tufo has an original identity that cannot be assimilated with any of the other grape varieties called Greek. It may seem strange, but this vine that everyone thinks is the son of a distant land, brought by some adventurous pioneer from the Peloponnese, is perhaps one of the oldest native vines in Italy. It is not an expression of a Greek culture but of an Etruscan culture and even today, if you look at how this vine is grown in the area of Tufo or Parete in the Caserta area, you will find many similarities with the cultivation methods that were used in the Po Valley by the Paleoliguri, which consists of entwining three or four shoots in a sort of braid to connect neighboring plants. This technique was called *rumpotinus* and was applied to those vines derived from the domestication of wild vines that had a high vigor, were exuberant due to the very confused, disordered vegetation, full of females. Of the wild vine, it also has round, whole, very particular leaves, bunches with a significant, unrepeatable shape: a main body, a very showy wing and a very evident stigmatic residue.
Greco di Tufo is genetically identical to Caserta Asprinio. They may have some phenotypic differences due to different anthropo-

genic selection processes, but their DNA is identical. These vines, up to 15 meters high, *a piede franco*, grown with poplar as a stake, in the past have always had fame and notoriety because they were destined for the production of a sparkling wine, and were also exported in the phylloxera period to France. That the system of vine cultivation on live stakes is an Etruscan heritage is also demonstrated by the presence, in the language of these people, of a word, *ataison*, referable to the Indo-European root and which means "to turn on itself", "to entwine itself", as the vine wed to the tree does. The expression "entwine" was also associated with vines (for example, Biblino) from very distant cultural areas, such as Asia Minor and eastern Sicily. The name of this vine, in fact, one of the first to be acclimatized in Magna Graecia, literally means "twisting vine" or "clinging vine", just as Aglianico had *anadendrite* as a synonym. The Greek agronomists of the 4th century described the vineyards of southern Italy as territories consisting of vines that supported themselves by living stakes, such as poplars and willows. This shows that the cultivation of vines on trees was not exclusive to the populations of central Italy and the Po valley, as it was also present in places that were also very distant from them. In fact, near areas where the spread of the two forms of training took place with a precise demarcation, in others, under the pressure of Greek colonization, the low-strain systems gave rise to expanded forms, such as for example in the production areas of Cecubo, on the northern borders of Campania, where the training system, originally supported by poplar, in the imperial age it had been replaced by the low stock, to reproduce the viticultural models of Greece to which the Latin culture referred, also for the use of slaves from Greece.
The sapling supported by a stake, even if it is now present only in old marginal vineyards, is itself an icon with an irreplaceable symbolic significance, an expression of choices made along a path of

centuries. It represents a fissure, a window through which we relate to a territory understood as an organized space, a place where structural components merge with those of form and character. The sapling therefore represents in its specificity, an object of culture, albeit in its composition with nature. That is, at least as long as culture is the result of a process along which a society finds its identity.
Most of these icons have disappeared. Those that have survived are only residual or consecrated (in museums) or marginalized, now out of place in the modern production contexts of Italian viticulture. Today's viticulture is the result of a journey of generations of winemakers where the succession of events has given rise to very complex realities so that it is no longer possible to recognize the evolutionary phases. The succession of processes of expansion and contraction as a result of social and economic crises, of edicts sometimes favorable and in other cases unfavorable to the cultivation of vines, of climate change and of major diseases, have led to the simplification of viticultural models (vines, forms of cultivation, cultivation techniques), where the need for effective mechanization discriminates against viticulture compatible with global competition, from that destined to disappear due to economic marginalization.

The metaphor of the journey and the narration of medieval viticulture

The journey is the metaphor of life. It includes overcoming difficulties, tests of knowledge, the search for the new, an escape from reality. It is a fundamental narrative tool starting from the *nostoi*, the return journeys of the Homeric epic of the Greek heroes from the Trojan War. Storytelling is the format with which the biographical path of the characters linked to the world of wine is realized, through the stories, the sagas, the tales, which are placed in a temporal dimension (in a historical context) and in a spatial dimension (the places of myth), where the often metaphorical, sentimental, inner journey, in search of the eternal Atlantis, who lives in legend, something beyond reality.

It is the adventure that generates its own story. The word "*aventure*" was born in the Gallo-Roman context and means "things that come to us, in an unexpected way". Medieval journeys on the paths of faith are adventures, just as the story of Italy's wine along the consular roads or pilgrimages is an adventure, so science is an adventure.

After the fall of the Roman Empire, the vine spread as a result, as so often in its history, of a god, a religion, as wine became a social symbol of European civilization.

At first, the spread of vine cultivation was an iterative, local phenomenon, due to the need to have wine for food consumption in small agricultural communities. The growing presence of small Benedictine monasteries then transformed viticulture into a fundamental collective phenomenon for the economy of entire regions, where the vine took on the role of a leading crop and its products were no longer only intended for local consumption, but also for export. These relationships between the many religious communities who had settled above all along the great routes of Christian pilgrimages, in a Europe still without borders, are the basis not only of the spread of those vines that had been created along the borders of the Roman Empire, but also of a new cycle of crossings, more aware than in the past, in the choice of parents, which sees Pinot and Traminer as the absolute protagonists, especially in those areas where the presence of Benedictine monasticism was greater.

This is an aspect of the social history of wine that marks a deep fault between the origin of the vines of continental Europe and the regions around the Alps, affected by the great roads connecting the south and the north, and those of the territories bordering the Mediterranean, in which the process of local domestication of the vine was less important and the cultural and genetic contribution of the Greek colonial experience prevailed.

The development of the paths of faith began in the 11th century, following a project for the transformation of economic and social structures resulting from the profound changes in early medieval agriculture. In particular, the connections with the maritime cities of Pisa and Venice allowed access to goods from the Levant, in particular textiles and spices.

The "via di monastery", the stretch of the Francigena that from the Po Valley crossed the Apennines to reach Tuscany and Rome and which had its focal point in Bobbio (in its monastery of San Columbanus the word Europe appeared for the first time), substantially followed that of the Roman road and connected a dense network of monasteries and hospitals. It was the heart of the *Regnum Longobardorum*, the *hidden limes* with the Byzantine territories located further east along the Via Emilia.

The ecclesiastical viticulture of the Carolingian era, starting from the ninth century, especially in the Po Valley, was flanked by an aristocratic viticulture, born by imitation, but which soon turned into a source of economic income. The spread of the vineyard was mainly in the suburban areas of the cities that radiated from them, but always a short distance from homes to avoid the frequent theft of grapes. Around the year 1000, bourgeois viticulture developed, as a result of a mercantile and artisanal class that had capital to invest in the countryside, especially in the production of wine, as a result of the exceptional population increase in many European cities. By ancient Greek and Latin custom, the grapes had to be pressed in the vineyard (*nullus habeat vel faciat mustum qui non habeat propriam vineam),* so along the roads were both the cellars and also the point of sale, even if this privilege was reserved for convents, while taverns could not normally be located within a radius of more than two miles of the cities to ensure that they avoided any municipal duties.

The ancient roads were not only a common thread between the Italian wine-growing areas, but a structuring element of the landscape. In particular, the Roman ones, with the centuriation, drew the rural space in orthogonal shapes and each centuria, consisting of three lots, was attributed to the veteran legionaries. This pattern traced by the *decumanus maximus* and the *cardus maximus* is still visible in the inter-farm road system in some areas of Romagna along the Via Emilia.

Medieval Italian viticulture through the lens of the Peutingerian table

Cartography is among the inventions that precedes writing, with the intention of charting the unknown and superimposing a structure on it. The cartographers of the early Middle Ages used maps like the pages of a book, with the intention of giving a theological vision of the Earth, capturing the imagination of those who consulted them.

This representation of the *eucomene*, a real storytelling *ante litteram,* of the world known at the time, entrenched Roman toponymy until the fifteenth century. Original cartographic tools, very different from globes, they consisted of regional maps and maritime charts, or pilot books, which appeared from the early Middle Ages to the fifteenth century, and date back to the Roman tradition, called "itinerari", whose characteristic was to trace a network of lines and not to represent a continuous surface.

The most illustrious of them was the Peutingerian table, a 13th century copy of a late Roman original of the mid-4th century (under the auspices of Theodosius, in 393 AD). Designed by Castorius, it takes its name from an official of the city of Augsburg, Konrad Peutinger, who received it in 1507 from the scholastic humanist Conrad Celtes. In the form of a scroll measuring 6.80 x 0.34 meters, the map illustrates the then known world, from Britain to India. In horizontal stripes, separated by the Mediterranean and the Adriatic Sea, mountains and rivers are indicated, as well as the names of the cities, represented by the drawing of small groups of houses, with small vignettes and the distances between them, calculated by the number of days needed to cross them. The main obstacles and how to avoid them are also mentioned.

The topographic layout forms a unit of communication with its immediate context. The figurative elements of the map are necessary for understanding the message it conveys. It's communicative intent, using graphics, is to tell tales of cities, convents, churches, castles, and on these traces and with the help of the testimonies of the Latin Georgics and chroniclers of the Renaissance, of vineyards, vines, wines.

The Table illustrates the backbone of a dense European road network. One itinerary could be preferred over another depending on the season, local conflicts, political conditions, or to avoid places infested by brigands.

In the section that conveys the Italian peninsula, in correspondence with the different localities it is possible to indicate, on the basis of Latin sources (Columella, Pliny, Virgil), medieval and Renaissance, the wines that were produced and the vines used. It is the first major route across Europe and redfines the idea of the road, distinguishing between "*via*" and "*strada*". In fact, the "*via*" indicate the way in which one proceeds, the "*strada*" instead includes the set of places that must be crossed to go from one place to another. We can also speak of "*vigneti di strada*", meaning by this definition the set of anthropological, political and social interrelationships that have marked the viticultural elements along the roads greatly travelled, often corresponding with cities of Roman origin (*oppidum, municipium, castrum*) or medieval, to understand whether in the configuration of these vineyards, especially in the cultivated vines, the religious hosts played an important role.

The current vineyards of Pavia, Piacenza, Lucca, San Gimignano and Viterbo are the legacy of those that had been established on the Via Francigena, as well as those of Ascoli Piceno on the Via Romea, just to remember a few of the most important staging posts for pilgrims on the way to Rome, and they preserve with their original topographical location and cultivated vines, the signs of that period.

In 1305 the Bolognese Pier de Crescenzi, in his *Liber ruralium commodorum*, was the first to describe, after the documentation of the Latin Georgics, about forty vines, placing them in many places in northern Italy between Piedmont and Romagna. In doing, so, he provided a precise indication of places where these vines were cultivated or in any case where the wines produced were made famous for their quality. Compared to the findings of the Latin Georgics and the Renaissance neo-Georgics, it offers a very precise description of the morphological and productive characteristics of the vines.

After de Crescenzi, in the early months of 1536, Sante Lancerio, wine merchant to Pope Paul III, wrote a guide to wines that accurately described in rich gastronomic detail the wines he encountered on his way along the Roman consular roads. In fact, he travelled along long stretches of the Via Cassia and Via Aurelia, from Rome to Nice, the Via Emilia and the Via Flaminia from Parma to Ancona.

Along the Via Francigena, and with its intersections with the salt and oil routes, medieval viticulture was born. Its structural model (of promiscuous viticulture, trees and vines) would last with until the arrival of American diseases.

The Metaphor of Anguish

Anguish is a feeling that has marked the last two thousand years of our history. The first age of anguish coincided with the fall of the Roman Empire, which took away the certainties of law and the protection from barbarian invasion. Christianity represented the only refuge from fear. The need for wine in the celebration of mass spread the cultivation of the vine across Europe through Benedictine monasticism.

With Probus's intervention to reverse Domitian's edict, the vine had returned to European villages. From those brought by the Romans from Pannonia (a province of the Roman Empire that included present-day Austria, Hungary, Slovenia and Croatia), the varieties that are still largely cultivated today were born.

By the turn of the fist millennium, "fake news" was spreading: the end of the world was nigh. In Europe, the desire to expiate sins drove Christians to the conquest of the Holy Sepulchre through the Crusades. As the Crusades in the Holy Land dragged on, in Assisi the Franciscan movement was born and the pilgrimages along the paths of faith that led to Rome became a global phenomenon that would be a catylyst for the development of the great sites of European viticulture (Champagne, Alsace, Burgundy, San Gimignano, etc.) along the Via Francigena.

The Via Francigena, in medieval iconography, is also represented as a labyrinth to express the anguish and sense of bewilderment that one felt in the face of a journey that might never be completed. The labyrinth is, for example, represented on a column of the cathedral of Lucca and in the church of San Pietro in Pontremoli, dating back to the 12th century.

The labyrinth is a single long and winding path to follow, which reveals itself only as you walk through it. It is like a network in which each point is connected with another so that those who travel have to continually correct the route they do not know and proceed by trial and error. The journey undertaken by those who enter a labyrinth and to pass through it with the risk of getting lost, searching for a way out, is an initiatory experience with deep religious significance. The wandering of the traveler, at the same time both inspired and confused, full of doubts and unexpected turns, is a metaphor for life, reaching the destination after a long journey to the "end of the world". The labyrinth is an archetypal structure, which reflects man's way of thinking, a sort of adaptation to the uncertainty of the future. But its maze is also a world of sin, from which the Christian must escape through the penance of pilgrimage

During the Fourth Crusade, Venice discovered in Monenvasia, a small port in the Peloponnese, a wine still unknown in the west, Malvasia, which would become fundamental for the upturn in the city's commercial fortunes that coincided with the Little Ice Age. This climate crisis, which lasted about four centuries, between the Middle Ages and the Renaissance, would dramatically mark the fate of Europe with plagues, famines and wars of religion, is at the center of the third age of anguish.

The many fears that still grip the world today, (such as COVID-19, which confronts us with the transience of life and its unpredictability, creating anguish), coupled with the effects of climate change and resource scarcity, are still present in the contemporary world. If the comfort of religion is now lacking, as in the past, the secular sentiment that sees our refuge in Mother Earth returns.

Ecological movements and the desire for sustainability have this ancestral origin. Protecting nature means giving continuity to our future. The ideals of environmental sustainability, which are associated with the effects of climate change, require a return to ecology, to Mother Earth, to less impactful viticulture (resistant vines, new more efficient rootstocks, cultivation techniques with less chemicals) and with less consumption of fossil fuels, production of renewable energy on the farm, based on the principles of the circular economy.

The Little Medieval Ice Age: A Lesson for the Future of European Viticulture

The centuries around the first millennium offered Europe a warm climate that pushed viticulture to very high latitudes, even as far as Scotland, and to equally unusual altitudes in the Alps, around 1200 meters above sea level. In this phase, also known as the Medieval Climate Optimum, viticulture expanded within the Alpine valleys, accompanied by olive growing, and occupied entire regions of continental Europe, thanks to the efforts of the monastic institutions. But suddenly everything changed.

Medieval iconography, especially in northern Europe, frequently uses themes that recall the living conditions of the rural populations of those times. It is enough to look at the vivid paintings by Pieter Bruegel the Elder (*The Hunters in the Snow,* 1565), or the numerous scenes of skeletons reported on cemeteries or sacred places depicting the so-called macabre dances or *imago mortis*, to understand what the climate, not only spiritual, was like in those times. The inclemency of the weather is testified to by a figure painted by Ambrogio Lorenzetti in Siena in 1338. The painting portrays a fully cloaked man on a gray background as dense snow falls all around and settles on the folds of his cloak. This way of representing the phenomena of nature was a real novelty in the 14th century, and the exceptional nature of the meteorological events of those years inspired his masterpiece *The Allegory of Good and Bad Government.* A precise reference to the climatic conditions of the Alpine regions towards the end of the 14th century is depicted in *The Cycle of the Months* at the Torre dell'Aquila at the Castello del Buonconsiglio in Trento. The famous fresco depicts the scene of the grape harvest in which some unusual but very effective details appear to reveal what viticulture was like at that time. The sapling training system, so dif-

ferent from pergolas but effective in keeping the grapes close to the ground, is striking, the covered press, to protect the must from the rain, and the clothing of the peasants, suitable for cold autumns.
So, what happened in Europe towards the end of the 14th century? Frosts, hailstorms, violent rains that lasted weeks, years of drought that caused severe food shortages, and the outbreak of pandemics such as the Black Death, which arrived from Constantinople with the ships of Genoese merchants and dramatized by Giovanni Boccaccio in *The Decameron.*
The age of Shakespeare coincided with a period of uncertainty and conflict (the European wars of religion), followed by the Thirty Years' War and the War of the Spanish Succession, as European populations suffered centuries of famine and starvation, culminating in the Irish potato famine of 1845 to 1852. An anomalous phenomenon, known by historians the Little Ice Age, which began with the Great Frost of 1709, destroyed European viticulture. It marked the beginning of the end of feudalism and the location of vineyards in Europe in this period underwent a dramatic transition. Viticulture disappeared from England and from all the inland valleys of the Alps. The moment became particularly critical because of the so-called "trade wars", which coincided with climate change, which made the production of low-grade wines unprofitable, as the new rates that were applied for the transport of wine were calculated on the value of the wine and no longer on the number of barrels transported. This caused the depopulation of many Alpine and Apennine areas, and therefore the loss of many vines that were the result of selection processes dating back hundreds of years.
The cooling represented a catastrophe for agriculture with the drastic reduction of cereals, fodder, fruit and grapes. The annual production of wine is an excellent indicator of the chronological succession of climatic events due to the habit, especially by monas-

tic orders, of marking the date of the beginning of the harvest, the current prices of wine and the quality of the grapes. As early as the 14th century, we have data on vineyards in Burgundy and southern Germany.

How did Europe's wine-growing areas react to the rapidly cooling climate? The first perception that something was changing was in the reduction of grape production due to the cold springs that hindered fruit set, and in the low levels of sugar content due to rainy autumns that did not allow the wine to be preserved the following year. Sometimes the low winter temperatures caused the death of the vines. The adaptation strategies varied according to the various wine-growing countries and the varieties cultivated, the forms of training and the different use of the energy resources of the territories, and the search for conditions of exposure of the vineyards that favored ripening. The first interventions usually concerned the form of training, with which they tried to put the bunch as close as possible to the ground so that it could use the heat it radiated. Forms of training on live stakes (tree-lined), an expression of primitive viticulture, were abandoned and low-stump (saplings) or low-pergola training systems, called "*a camera*", were adopted. Viticulture was moved to hilly positions, otherwise used for grazing because of their low fertility. But this choice favored above all the great noble and ecclesiastical estates, which had large areas of land and were thus able to control the entire market of quality wines. Only the wines of the eastern Mediterranean, marketed by Venice, were able to counter this monopoly, which for some wine-growing areas such as Burgundy, Alsace or the Rhineland, would end only with the fall of the *ancien règime.*

Genetic innovation underpinning the response to climate change

The first reaction of European winemakers to climate change was to favor the early-ripening and white grape varieties. An exemplary testimony to this is the trend in grape production per hectare from the 16th to the 19th century around Lake Caldaro in southern Tyrol, recorded by the Chapter of the Cathedral of Trento, owner of vast vineyards in that locality. Two contrasting trends emerged: the first, coinciding with the most acute climate crisis between the end of the 15th century and the beginning of the 16th, during which the areas of Vernatsch (red grape variety) were reduced and those of white grape varieties (probably consisting of numerous varieties) grew, and the second, in period after the 18th century, in which this trend was reversed and the production of Schiava, a very generous grape, resumed, while those of white grapes were reduced.

However, the choice of new varieties proved to be decisive, implemented both by selecting the seedlings that developed from spontaneous crosses that showed particular qualities of precocity in the ripening of the grapes, and by introducing vines from other areas, less favored climatically. This is the case of Champagne, which abandoned red grape varieties such as Pinot Noir and extended the cultivation of Gouais, which had arrived with the legions of Probus and which had contributed to the creation of Chardonnay, in a cross with an ancestral Pinot.

In Veneto, in particular in the Treviso area, especially in the later stages of the cold thermal cycle, coinciding with the Great Frost of 1709, came the abandonment of the late ripening varieties, often the best. The return to more favorable climatic conditions, saw a return to the planting of more productive but lesser quality vines to meet the strong demand for wine from the European market.

Thus, vines such as Marzemino, Bianchetta, Malvasia, Piccolit disappeared, replaced by the more neutral and acidic Verdiso, Rabbiosa (Raboso), Dall'Occhio, etc. Despite improvements in production methods, the quality of the wines, especially in France and Germany, was very low, and for this reason they were intended for popular consumption. Nobles and high clergy, unsatisfied with such wines, looked elsewhere for alternatives.

Venice and the metaphor of innovation and competition

Paraphrasing one of Voltaire's best-known expressions, *l'esprit des nations*, derived from the *Assai*, his essay on universal history, the spirit of nations defines the structural characteristics, the centuries-old and long-lasting legacies that give a territory an unmistakable identity, almost a genetic code. In creating this spirit of nations in the *Tre Venezie* (the three historical regions of Venezia Euganea, Venezia Giulia and Venezia Tridentina), the region's rich winemaking tradition made an essential contribution.

There is a red thread that binds the countless pieces of the mosaic represented by the territories of the Veneto, which can be well summarized by the early and intense mercantile activity linked to wine represented by the Greek colonial experience, which began with the interest that the Etruscans showed in the trade of the emporiums of Adria and Spina. Even today it is possible to detect in the loanwords, the passage of many terms related to viticulture and oenology, from Greek to Latin, through Etruscan, to testify how the cultivation of the vine had an indigenous basis, while the Greek-Oriental contribution was decisive in the refined civilization of drinking. It cannot be denied that another unifying element in the wine-growing history of the Veneto region was the presence of the Republic of Venice, in its territories from the lands of Istria to the Vallagarina of Trentino. The image of the lion, the so-called war lion, which stands out on the old entrance gates of the cities occupied by the Serenissima or in the frescoes on the most important palaces, is an indelible sign of a culture that has pervaded the life of the populations of the Veneto through a language that with a great sense of belonging made them feel like the protagonists of European art, culture and civilization of the time. Venice taught these peoples not so much to produce something (the Venetians were known in the Middle Ages as people who "*non arat, non seminat, non vindemiat*" [don't plough, don't sow, don't harvest] as the anonymous *Honorantie civitatis Papie* quoted), but to open up to international trade, distancing themselves from the bigotry of the princes, the Bishop of Tyrol and the decadence of the Duchy of Friuli.

Venetian merchants were the heirs not only of the Etruscans and Phocian Greeks who had first inhabited those places on the shores of the Adriatic, but also of the most astute Levantine merchants, from whom they had learned the refined techniques of trading. The long history of wine trade reached its climax with the impact of the Little Ice Age on the production of wine in Europe between the mid-14th and 18th century. This allowed Venetian trade to control almost the entire quality wine market from the eastern Mediterranean to the rich regions of the North Atlantic, allowing the birth and development of a so-called "*di imitazione*" viticulture in Italy for the production of Vinsanti and Malvasia to replace Mediterranean production.

It was an indication of the commercial strategies of wine in the following centuries, as the maritime cities of Venice and Genoa marketed wines from the eastern Mediterranean (Malvasia and Vinsanti) or from the regions of the French Midi and the Spanish Levant (Vernacce), using not the name of the vines but the more or less mangled place of production (Monemvasia, Santorini, Vernazza, etc.), offering a few wines with very precise organoleptic characteristics, easy to recognize and, due to the high alcohol content, transportable without undergoing alterations in flavor.

The competition between the wine merchants who belonged to the different Italian maritime republics had distant roots. It concerned sweet, aromatic and alcoholic wines (Malvasia and Vinsanto) produced in the eastern Mediterranean and marketed by the Venetians and partly by the Genoese (the Ligurian Vernaccia), on the Anglo-Saxon markets, and the sweet wines obtained in Atlantic and German viticulture due to the action of noble rot, also destined for the rich markets of northern Europe, but made known above all by Dutch merchants who were precluded from entering the Mediterranean.

In retracing the history of a city, the thought instinctively goes to the origins, to the events that in the course of its history have determined its fame. The sands of time, however, are inexorable, memories become more and more blurred, and the reconstruction of the most distant facts must resort to clichés, to sensations that gradually take on the contours of myth, of legend. One of these identity myths, perhaps the best known, on which Venice's fortunes were based was the one linked to wine. To describe this history it is necessary, however, not to limit oneself to a treatment of events, of battles of famous people, but to transform history into anthropology, restoring Venice to its original color, referring to the physical and social reality of everyday life where ideas are born. The myth of Venice, therefore, is unthinkable outside the historical-social framework constituted by the Republic, a city-state, characterized by the assembly management of power. These considerations make the Venetians of that time our contemporaries, allowing us to recognize ourselves in them, but on the other hand it distances us from them, makes them resemble the Levantine societies from which they have assimilated so much.

Two complementary ways for measuring similarities and differences between current humanity and that of the past. If this brings us closer to the ancient inhabitants of the Lagoon, it also leads us to discover our otherness, that dark background that makes us strangers to ourselves, but through which the enigma of the history of Venice allows us to enter the paradigm of modern man.

Venice is not the easternmost of the great western cities: it is an eastern city. It is enough to observe the golden domes of St. Mark's Basilica at sunset to be convinced of this. Even the conditions in which Venice took shape during the early Middle Ages favored the consolidation of forms of mentality and particular civilizations that originated from the special bond it had with Byzantium, and that made the city a hinge point between the great cultures of the time. Early medieval Venice was different to the West of the feuds and castles for its eminently and precociously mercantile economic structure. On the trade routes, however, not only goods and money traveled, but also and above all the ideas that contributed to forming peculiar attitudes and mentalities that found an original confirmation in the sense of the State, in that idea of the common good advantageous to all, from the patricians to the commoners. Venice seized the opportunities offered by some particular aspects of that phase of the Middle Ages: the development of the cities (where wine consumption was concentrated), especially in Northern Europe, the need, in a market economy, to supply foodstuffs from distant territories; and the growing demand for wine, a particular type of wine, among the emerging social elites (such as merchants and the bourgeoisie), stimulated by its preciousness and the consequent high cost that increased its symbolic value. The theme of the "commercial revolution" interpreted in a Mediterranean function, can be linked to the Italian myth of the Maritime Republics, only if it is related to the history of the cities of reference, as places of exchange and trade with the territories with which they interact.

Since the 10th century, Venice has linked the Adriatic area, the Balkan area, the Greek world and Egypt with the rich regional hinterland that places it in constant and profitable relationship with the

German world. It was the global market, the network of its trade, that made Venice, a city no less exotic in terms of modernity and contamination of customs than its Chinese or Indian counterparts, the port in which all the commercial powers of Europe would stop in the following centuries.
From the 14th to the 16th century, European society was very careful to define the lifestyles of different social groups. Food and wine choices were a privileged tool of the elites to mark social superiority. Wine, in particular, offered an important element to codify the social distinction. For each social class there was a type of wine. The sweet wines of Cyprus and the so-called Greek wines from the places that coincided with the ports of call of the Crusades, occupied by the Venetians in the transport of soldiers during the Fourth Crusade, or the Grenache, Garnacha or Vernacha wines, which designated both a grape variety and the homonymous wine of Mediterranean style, belonged to the types of sweet wines, liqueur, high in alcohol, that were the prerogative of the wealthy classes. At that time there was a change in the choice of wine color. As for the bread intended for the lords, which must be white, made of wheat, and the black one of rye or spelt, was for the peasants and servants, the red or black wine was intended for the "working people", while the white wines and the Clarets were for the people who "live in idleness". In the choice of color, the effect of climate change caused by the Little Ice Age cannot be ignored, which hindered the ripening of red grapes, and at the same time the arrival in Europe of "Greek wines" of white and amber color.
The primacy of sweet wines in the Middle Ages was indisputable and usually referred to white wines, imitation of wines that came from Greece. The innovation represented by the bottle is not limited to extending the life of the wine to facilitate its transportation, but also to new trading possibilities represented by the improved

storage capacity, which allows you to control not only its evolution over time but also to avoid price drops due to the need to sell the wine prematurely due to its precarious conditions of stability and durability.

Once again, product and process innovation was accompanied by the availability of economic resources that allowed these wines to become symbols of wealth and distinction, and therefore to be consumed by the wealthier classes. It is curious how the bottle immediately became a status symbol, and not only because it was a convenient container. Venice, about a hundred years earlier, with the development of the art of glassmaking in Murano, produced not only heavy glass wine bottles, but a series of accessories to serve it, of clear glass and of great stylistic finesse, which allowed for the first time, to the consumer accustomed to pewter, silver or tin containers, to see the color of the wine he was drinking in transparency. This innovation, which today appears almost ridiculous, is testified, for example, in Veronese's painting of 1563, *The Wedding at Cana*.

From that moment on, wines, especially sweet wines served at important tables, were served from transparent carafes defined as the Venetian style. Of magnificent workmanship, the wine had to be clear and therefore produced with more refined oenological techniques than in the past. In the rigid ceremonial that distinguished the official banquets of *La Dominante* between the 16th and 17th centuries, where the dishes served were a melting pot of the gastronomic culture of the Mediterranean, between sweet and sour flavors of Byzantine inspiration, meats and fish ennobled by the fragrances of the most disparate spices, the dining-room guild, which was entrusted with the task of organizing the sequence of dishes, began every table service in the fashion of the viziers of Constantinople, with royal pastries (marzipan) accompanied by sweet Marzemino.

For the King Henry III of France, the Hall of the Great Council opened and the menu for the occasion consisted of dozens and dozens of dishes. The king, frugal and inhibited by the fear of poisons, sipped only a glass of sweet Marzemino. This was the homage that Venice paid to the sources of its wealth: to those wines, to those spices, to the sugar of Cyprus, to the salt and to the thousands of other goods that came from all over the Mediterranean to its warehouses and that testified to its strength and wealth on the markets of rich Northern Europe. The myth of the sweet Marzemino remained until the end of the independence of the Republic, which took place on 17 October 1797 and ironically became the symbol no longer of power, but of decadence.

It was the Venice of 17 theaters, an essential stop on the Grand Tour, identified in the 18th century as Europe's capital of frivolity, unparalleled scenographic art, baroque music and informal costumes. A setting in which Mozart composes his *Don Giovanni*, in which Emanuele Conegliano, alias Lorenzo da Ponte, after the abjuration of the Jewish religion, creates the dual characters of Don Giovanni and his servant Leporello: debauchery and conscience. Eclectic and a lover of the good life, a regular visitor to salons and gaming halls, da Ponte could not fail to mention in his libretto of Don Giovanni "the excellent Marzemino", a wine fashionable in the Venetian aristocracy, together with Tokaj. Most likely, da Ponte remembered the Marzemino that the Collalto counts, ambassadors of the Republic, to the Empire, served in their palace in Vienna in Platz am Hof, where Mozart made his debut with his sister on 9 October 1762.

Perhaps few people know that Venice lost the market for sweet wines in northern Europe, one step away from the creation of the "great wines", with the cutting of batches from very different locations, because it underestimated, on the threshold of the "beverage revolution", the role that the "spirit of wine", the alcohol produced by the distillation of wine, played in the creation of fortified wines and mutated musts that had a much cheaper price than those imported from the Mediterranean. Venice also made another mistake, because it responded to wines produced using fortification, in particular Port, which relied on English capital for production and trade, imitating a product then fashionable on the French market, absinthe, with the production of vermouth. Paradoxically, this change in style to Malvasia, which had been closely identified with the Venetian trade, turned out to be a counterproductive innovation, as the consumer associated it with wine counterfeiting.

For about a hundred years, the market for fortified wines passed into the hands of the Dutch, thanks to their great maritime power, who then had to cede it to the English on the threshold of a beverage revolution because of the protectionist policy of the latter, with which they replaced French wines (such as, for example, the strong muscats of the Midi) with Spanish and Portuguese wines. They too practiced the mixing of alcoholic products of various kinds, as a fundamental aspect of their work. But the risk of being pointed out as sophisticated was foiled by two great innovations, one represented by the discoveries of chemistry and microbiology that made it possible to produce very pleasant and defect-free wines, but above all by the consumer who was moving from Claret wines from Bordeaux to other drinks with very different aromas to wines.

Italy and wine, an idea of the future

This constantly changing world makes cultural ties uncertain and causes people to question their roots, forcing them to adopt other people's rules. To prevent standardization from leading to the disappearance of the original cultures, it is necessary to invent new cultural forms. This is what happens when two diverse cultures meet. Indeed, the most interesting things often happen on the margins.

Italy is the result of many cultural contributions from every corner of Europe. Germans, French, Austrians and Spaniards, to name but a few. Over time, with the expansion and mixing of original cultures from different systems, the periphery has become the center. It is the result of contributions of vines and viticultural techniques with distant and mysterious origins, which taken individually appear unfinished, and only in this viticulture do they find a perfect synthesis, a fusion of products, stylistic features and languages that satisfies the desires of eclecticism and syncretism of the modern consumer. Eclectic is the one who moves with ease combining different styles. Syncretic because he manages to blend them harmoniously, creating a seductive synthesis.

If the main resource of tradition is its ability to protect and hoard diversity, it is precisely in the uniformity of the uncritical transmission of the past that traditions die. To be able to give our viticulture a new future, we must not rely on traditionalism or technological innovation, but start from a correct interpretation of tradition by focusing our attention on a crucial node of the wine supply chain: the communication of the values of our viticulture.

The birth of oenological science in the 19th century coincides with the therapeutic action of the scientist against the numerous "wine diseases" and consequently on the gustatory defects that these

cause. So, typicality linked to territory, a characteristic of belonging to Italian wine, to a particular category, is a consequence not only of a geographical origin but of human factors and claimed by society, bearers of memories.
How then can we give its original meaning back to the word typicality, which makes an Italian wine almost readily identifiable from its chromatic or typological characteristics, as we do when we recognize a person by the salient features of his face or a beautiful painting?
If we associate the meaning of some expressions such as originality, identity, naturalness, tradition and typicality (the latter in the Weberian sense), values that are the basis of a quality agri-food product, to the perception that the consumer has today of DOC wines, we realize that this relationship is only semantic, not substantive. The causes of this fracture are many, but what has played a decisive role in the failure of the vast majority of Italian denominations has been the absence of a culture of the territory of the producer. Such a culture develops in the relationship between the characteristics of the environment and those of the grape variety.
The production of wine has thus been interpreted as a demiurgic exercise, where technology had the fundamental role and the grape, with its compositional characteristics, was only a matrix to be corrected in various ways to obtain wines that met the changing needs of the market. At the base of a production process such as that of wine, so closely linked to the civilization of the Mediterranean populations, there must be an ethical principle with which perhaps the legislator 50 years ago hoped to give a soul to the law of DOCs, but which unfortunately has not transpired.
If we abstract from the topographical, geographical expression of a wine-growing territory and observe it for the deep meanings it emanates, it appears to us as a tablet, where the *logos* (the word,

the thought) materializes in a wine. A tablet like the one where the commandments that Moses offered to the Jewish people were inscribed, like the altar of sacrifice where the pact between man and god is sanctioned. The wine-growing territory is just such a table. A place where the alliance between the producer and the consumer is realized in compliance with the rules of the denomination and the increasingly fragile environmental resources.
Respect for the land means asking the land for what it can give us. This attitude translates work into "an alliance with the land". The perspective of respect, however, presupposes, first of all, a thorough knowledge of what one respects. Saint Augustine affirmed that one cannot love what one does not know. The studies of viticultural *terroirs*, addressed in recent years with methodological approaches of increasing complexity, in an attempt to grasp the secret that binds the sensory characteristics of a wine to the origin of the soil where the vine is grown, have certainly contributed to removing some important aspects of the interaction between the vine and the environment that limited the vegetative-productive expression of the varieties, which are the basis of the fine chemical composition of a wine. Such studies try to make the technical arguments prevail over the legal ones of the territorial delimitations, but above all they emphasize the increasingly important role of the consumer in the purchasing decisions of a wine, not only for its organoleptic contents, but also for the symbolic values linked to the place of production.
Viticultural zoning has become a tool that mediates the emotions aroused by a landscape and seeks the synthesis between culture and nature, in order to offer way to save symbolic representations and environmental needs, to connect aesthetic instances and economic facts, to respect production tensions and tourist needs.
The origins, the roots, in themselves are not of much help. They

are a necessary but not sufficient condition. Marc Bloch, one of the *"maître à penser"* of the 20th century, explained this statement with an example: without the acorn, the oak would not exist, but not all acorns become oaks, and if they become so it is only because in that climate and in that soil the acorn can express its true potential. This potential, which some vines express only in very specific areas of Italy, is not only represented by the characteristics of the soil and climate, but above all by the know-how of its winemakers. But something important is still missing. There is a lack of awareness that wine is, for Italian winemakers, "a common good", like water, school and the environment. That's why viticulture must be supported by a social pact. We cannot live well if we do not feel first and foremost part of a common destiny and that a greater *esprit de corps* and the pride of belonging to a land so rich in history and natural resources, overcomes localism and jealousy.

At the basis of this complex phenomenology there is the polarization of Italian viticulture which has occurred in recent years as a result of the globalization of international markets. From a situation of great diversity, represented by a multiplicity of small denominations where native vines were the crucial element of a wine, to a viticulture imitating the French one, represented by five or six denominations well-known internationally, which alone represent more than 60% of Italian wine exports.

This opacity in consumer choices has caused disaffection especially among older winegrowers, whose income has progressively eroded and which has led to the transfer of planting authorizations from the southern regions to the rich viticulture of the north-east, where alongside established native vines such as Glera or Corvina, Pinot Grigio and vines for the preparation of sparkling wines are constantly expanding. In this economic and social scenario, university-led studies have also been lacking into areas such as native viticultural heritage and viticultural research, cultivation models adapted to ancient vines, the control of production / vine, the dynamics of the ripening processes, and the effects of climate change and how it can be used to enhance the compositional characteristics (tannins, acidity, color stability) of certain vines previously deemed unsuitable.

Facing the future by knowing the past

If we do not know the past, we cannot prepare for the future. Today we are in such a hurry, but if we had the patience to reconsider the events that have shaped European wine, we would find invaluable insights to better plan for the future.

With the so-called "beverage revolution" at the end of the 17th century, the consumption of wine, spirits and the spread of non-alcoholic beverages of non-European origin, such as coffee, tea and chocolate, underwent a profound transformation that caused a crisis in some of France's most famous wine-growing regions. In particular, in the Bordeaux area, the Clarets destined for the English market were replaced by the darker wines of Cahors. Similarly, viticulture developed in territories unsuitable for producing quality wine, but which instead proved favorable for wines to be distilled (Cognac and Armagnac), also benefiting from their proximity to the Atlantic ports of embarkation to the northern countries. But the real novelty was the emergence of new wines, such as Champagne and Port.

Venice reacted to the wines produced by distillation, in particular Port, by imitating a product then in fashion on the French market, Absinthe, with the production of Vermouth. Paradoxically, this change in the type of a drink, Malvasia, which had been identified with the Venetian trade, turned out to be a counterproductive innovation, as the consumer perceived it as a synonym for wine counterfeiting.

What were the causes of these changes? Amongst the most significant factors were climatic and demographic events: the climate crisis of the Middle Ages that resulted in the Great Frost of 1709, the arrival of the Black Death in Venice and Genoa in 1348, the wars of religion, and the social consequences of the Protestant

Reformation. Of lesser influence, we can include various phenomena such as the arrival in Europe of American diseases (phylloxera, downy mildew and powdery mildew), the changes in the geographical distribution of viticulture caused by the increase in wine transport rates in the Middle Ages, the disappearance of high-altitude viticulture in the Alps, the economic and social changes underlying the conversion of viticulture from promiscuous to specialized and vice versa, the effects of wars or military invasions, cultural stimuli or economic stimuli from other countries, especially in border areas, the circulation of wines from other wine-growing areas and changing food fashions.

The so-called phase of "permanent innovation" began for the wine industry towards the end of the 17th century. Three factors can be identified that contributed to this change: knowledge, competition and capital, which applied above all to the production of fine wines for aging and sparkling wines.

Knowledge refers to developments in the chemistry of fermentation, the identification of fermentation agents in yeasts, the use of sulfur dioxide as an antiseptic and preservative, and the industrial production of cork bottles and stoppers.

Capital is a necessary condition for those who produce and buy wines for aging, both to endure the time the wine spend in the cellar, and to pay for bottles and corks before an enterprise reaches profitability.

In the latter half of the 20th century, the French Oenologist Émile Peynaud inspired a change of perspective that transformed enology into an aesthetic practice aimed at reducing the effects of randomness on the taste of wine. Today we do not drink the same wine that we drank only a few decades ago. The invention of winemaking tradition and orthodoxy is a trait of today's popular culture that has no historical foundations. The greatest risk for this important sector is that of falling victim to its own rhetoric, while a realistic view of consumption behaviors and trends is essential to consolidate the economic assumptions of the sector.

At the turn of the 19th century, the awareness that living conditions had improved due to the results obtained by scientific research establishes the perception of the importance of science to the progress of humanity. The use of science solves any problem and provides prosperity unthinkable in the past. This optimistic sentiment constituted a powerful new incentive in the birth of forms of collaboration between science, represented by the university, and industry, especially with that of chemistry and the applications of electrostatics.

What sociologists call a "socio-cultural" space, where everyone's experience contributes to the construction of a virtual dimension but which has the connotations of a topographic map where the places (of the mind), the trajectories (the ideal paths), the destinations (pleasure) are indicated. As in Kant's experience, our mental (topographical) map is therefore identified with temporal and spatial variables, which allow us to decline quality through three shades: seniority (the immanent quality of age), authenticity (the quality certified by tradition) and stability (the quality acquired for duration).

The decline of an economy always has to do with its inability to adapt to changes in the market, in terms of consumer tastes and production methods. This had happened in Venice only in the seventeenth century, when it had to face the new competitors, the English and the Dutch, who were able to adapt promptly to the purchasing power and tastes of the new clientele, as well as take advantage of the loss of its ports in the Mediterranean. There was also an attachment to tradition, which manifested itself in the rejection of the new "Atlantic" wines, which were instead in demand by the rich market of the European bourgeoisie. If we do not know the past, we cannot prepare for the future.

Wine storytelling and young people: an "oenosophy" for responsible drinking

A recent survey by the HBSC (a collaborative transnational study on the health and well-being of adolescents) showed that 29% of young people around the age of 15 consume alcohol at least once a week, though rarely wine, more frequently the so-called *alcolpops*, a mix of fruit juices and spirits, much more subtle towards dependence. Drinking is "cool" and helps a young person to overcome the enormous problems of mistrust and disesteem of puberty, a sort of rite of passage. Wine is therefore consumed to overcome, through *binge drinking*, the typical transgression of adolescence.

In this period, "emotions illiteracy" leads to a deficit in understanding the changes that are taking place in an increasingly "fluid" society, where too much information circulates, and it becomes difficult for the recipient to extract meaning that has an actual value.

Start with young people. Who better for a course in the culture and civilization of wine starting from primary school? But such an approach must also involve the professional training of restaurant workers. Of course, a proper training plan is needed to implement such a project.

In Italy, in the 1990s, in response to the increasingly pressing problem of alcohol-related issues in young people and the need for early intervention of training and food education in general, an association was founded, The Permanent Observatory on Youth and Alcohol (OPGA). A community of researchers, experts and communicators, the observatory made its expertise available to provide information on the issues of the relationship between young people and alcohol, making the case for an anti-prohibitionist and damage minimization approach. Working with teachers, the "Laboratory on the behaviors and effects of alcohol consumption among young people", which was a product of the Observatory, established at the University of Perugia, recently launched a project to prevent dangerous consumption of alcohol.

These initiatives, which are very worthy in terms of intentions, in addition to representing a valid starting point, have however failed to have a widespread and pervasive impact on Italian schools as an *ad hoc* law might do. Italy is not, therefore, at "year zero" in this field, but in order not to diminish the work done so far, more incisive and coordinated initiatives are needed.

The French poet Charles Baudelaire wrote: "Wine is similar to man: it will never be known to what extent it can be esteemed and despised, loved or hated, nor how many sublime actions or monstrous deeds it is capable of." Only culture and knowledge can make a difference.

When at the beginning of the 1980s, Masquelier, professor at the University of Montpellier, stated that the alcohol of wine is like a tree that hides the forest, meaning by this, the cliché that identifies wine in its main component (after water), ethanol, and thus prevents the enhancement of the health properties of hundreds of compounds present in it. At that time, French public opinion was against the dangers of alcohol abuse, considered by the majority of doctors, especially gastroenterologists and hepatologists, to be responsible for serious dysmetabolism. Even urban sociologists saw the early consumption of wine in young people as a social phenomenon that had serious repercussions on delinquency and school dropouts. France had also known the problems of alcoholism since the end of the 19th century, with the spread of cereal and wine distillates, coinciding with the so-called industrial revolution, which had transformed entire generations of malnourished peasants into tired and frustrated workers, who found in alcohol the only comfort for their despair.

In this period the first anti-alcohol movements were born, also backed by the socialist ideology of the time which was strongly opposed to the consumption of alcohol by the proletariat. Unfortunately, this assimilation of wine to alcohol has prevailed over that of wine as a healthy drink up to the present day. It was only the extraordinary media impact of the "French paradox" on the American television audience at the end of the 1980s, especially in relation to red wine, that wine has returned to its important place in the modern diet, at the center of the "Mediterranean" lifestyle renowned for its beneficial effects on the incidence of cardiovascular diseases. This lifestyle and the habit of drinking wine at the table with the family, was also responsible for the lower severity of the phenomenon of alcoholism in Italian and Mediterranean populations in general, compared to those of northern Europe. Wine was a complement to food and its sharing between young people and adults diminished its association with transgression.

Introduced in France in 1991, the *loi Évin* banned all forms of advertising for wines, spirits and tobacco. The main objective of the legislation was to transform wine from an exciting beverage, into an instrument of national culture, starting from primary and secondary school education. To this end, it allocated important resources in the organization of traveling exhibitions that educated French students, through the description of the production *terroirs* of the most important appellations, grape varieties and cultivation techniques, winemaking and distillation, and finally, in an interactive way, contributed effectively to the knowledge of wines through their aroma, with aerosol diffusers that reproduced their sensory profiles. These initiatives have on the one hand reduced the negative impact of alcohol on young people and on the other contributed to increasing the pride of belonging (if there was any need!) to a country that has always made wine an important tool to communicate its best aspects. Recently, through a new bill presented to the National Assembly with the support of the Vin et Société association consideration is being given to replacing this now obsolete tool of communication, with the establishment of a television network dedicated exclusively to wine, to the dissemination of its knowledge, to the discussion of the problems of its production and consumption.

Telling young people about wine

The values, the emotions of young people and wine

Wine storytelling for young people must reflect the anthropological differences between generations:

Baby boomers: (no longer young), born between 1946 and 1964, children of the post-war population explosion and the economic miracle, today's boomers represent a large and influential component of the active population (Bill Gates, born in 1955, one of the richest people in America; Donald Trump, its president, born in 1946). The only substantial difference from their parents' standards, is that they are much more sensitive to quality of life. Physical health comes next, from jogging to aerobics, natural foods to vitamins. The BMW is their preferred status symbol.

Generation X: born between 1965 and 1980, they have lived in a world that is obsessed to levels of collective paranoia with values such as peace, environmentalism, a return to traditional values and optimism. Now in their sixties, they are overeducated, underemployed, dismayed, lazy nihilists with a sense of humor. They were the first generation to not know war, to follow the social revolution of '68 and feminism, and to reach higher education en masse. They grew up in a world with the energy crisis, AIDS and recession. They are the children of divorce, parked in front of the TV. Their world is inspired by the philosophy of "new traditionalism", that is, a return to traditional values but safeguarding social, economic and technological achievements. In *A Hologram for the King*, the journey of a failed entrepreneur who embarks on a journey to the desert to save himself, the story mirrors that generation without an identity as defined by Douglas Coupland. They need actions to feel alive, to finalize an individual experience.

Millennials: born between 1981 and 2000. Erasmus students, free from compulsory military service, well educated, desperately unemployed, they emigrate en masse like their great-grandparents. Free from the past to allow themselves to regret it (and to mix it up, reinvent it, contaminate it), putting together the nostalgia of the old with the commodities of the new (ancient vines and biodynamics, amphora and web). A ready made of passions and obsessions, beliefs, trends, utopias and nostalgia continuously updated. Social media savvy, interested in culture, disenchanted but worried about the future. For them, food is synonymous with identity and a fundamental source of well-being, favour for fresh foods if not organic, it is not only an organoleptic experience but also of culture and places of production, They practice food and wine tourism.

Generation Z: born after 2000 (because they have reached the end of the old millennium), their life is defined by the "i" (iPad, iPod, iPhone), they have digital in their DNA, a great ability to interact with other cultures (thanks also to their proximity to second-generation immigrants). Giorgio Ghiotti, a young Millennial writer, born in 1994, tells in his book *Dio giocava a pallone* ("God played football"), stories of first loves and first transgressions of the teenagers of Generation Z: "When I go to schools, to talk about books, I see teenagers hungry for emotions, exactly as we were at their age. They listen, but they don't want directions, nor already traced paths. Generation Z walks with its own rules and adults can only adapt."

The actualization of the myth of wine in young people: a path of knowledge and conquest

Knowledge and joy

The original goodness: Wine is intrinsically good. The first drops given by the grandfather, a remedy against colds, to celebrate the loss of the first tooth, so it is a family drink.

The transgression: Wine is a drink of adults. It becomes a right of passage, a symbol of coming of age: first beer, then aperitifs (with sparkling wine and spritz), special occasions, school trips, summer parties, birthdays, getting tipsy, the first drunkenness.

Wisdom and passion

The death of Dionysius: his myth is something for intellectuals, elitists. Young people identify him with their parents, with their knowledge of wines, the search for true and genuine wines (natural and biodynamic wines), anti-snobbish, represents independence.

The relationship: growing up means leaving the group as Jack Frusciante (*Jack Frusciante è uscito dal gruppo*) teaches us. But leaving the group means conquering a new level of relational skills, choosing new friendships, deepening communication. Wine ceases to be a transgressive element, to become a disinhibiting factor, to reduce distances. It is, however, necessary to know the different wines, as one does for food, almost to defend oneself from the bad part of wine, that of alcohol. Knowing is a form of control and a way to express oneself, the sign of the transition to the adult world

Disinhibition and control

Care as a return to the origins: the search for truth, making visible what is invisible through experience and knowledge. Coming out of adolescence, wine becomes a vehicle of expression, of taking care of oneself and one's relationships. It becomes an expressive means to connect the idealized attraction of childhood with the disappointing impact of adolescence, to the awareness of being adults, but also to a healthy product linked to the earth and its origins.

Territory and tradition

The land is the label: good wine is that of one's origins, but not the Australian or Californian wine (they have no traditions, they do not know how to associate it with food); that of well-defined territories, with specific territorial characteristics (knowledge of the geological origin of terroirs), All wines identified by their geographical origin are good. This is where the price range and the label come into play (different *terroirs*), different quality of wines based also on their fame, for example Sassicaia, or Tignanello) and the space for communication opens up.

The eternal return: from myth to social media (narrative marketing). Myth is never a question of denomination, of proper names, but of the effects that underlie them. Dionysius and the protagonists of the myth are for young people and communicators, old ironmongers, but the original elements that act behind them are not; it is the path that Dionysus takes on the path of emancipation, as young people do.

The path of wine as an original good, which disappears in transgression, which re-emerges as a conscious choice, what is it but the path of the god who immerses himself with nature in the underworld to re-emerge in spring with the return of life. And isn't Dionysius perhaps the god of emancipation and liberation on a path that is retraced even today? To use this myth, however, it is necessary to start from some concepts that underlie the original myth of wine: the identification of the earth goddess with the mother vine, the vine goddess, from which the origin/earth pair is born. Wine is born from the union of wet and dry, of air with earth, of soul with spirit, of wisdom with passion. The archetypes to be evoked are: wisdom and passion for balanced consumption (ambivalence of the wisdom and passion of the Greek world); wine as a cultural marker, knowledge and joy for mature consumption, as knowledge of oneself and others for a better relationship that gives pleasure (wine, the blood of Dionysius, the symposium, belonging) disinhibition / control, territory and tradition (typicality, belonging).

L'ARTE DEL WINE STORYTELLING

Prefazione di Stevie Kim

Non è un caso che la pubblicazione di questo libro coincida con l'ottantesimo compleanno del Professor Attilio Scienza. Un omaggio ben studiato? Forse. Una scusa sfacciata per un altro giro di festeggiamenti? Decisamente sì. Non che Scienza abbia bisogno di pretesti: ha più energia, dinamismo e forza intellettuale di persone con la metà dei suoi anni, il che è, a dir poco, ingiusto. Posso confermare in prima persona quanto il mondo del vino possa essere fisicamente - e mentalmente - sfiancante; eppure, mentre molti nella sua posizione si godrebbero una meritata pensione sorseggiando un grande vino in qualche vigneto baciato dal sole, Scienza rimane una delle voci più instancabili e richieste della scena vinicola italiana. Quest'uomo è un'enciclopedia vivente, parlante, insegnante e, occasionalmente, anche brontolona (se solo le enciclopedie avessero un debole per le digressioni aristoteliche e le analogie omeriche).
Sebbene non abbia un account sui social media, chiariamo subito una cosa: Attilio Scienza non li snobba affatto. Anzi, l'ho personalmente "arruolato" più di una volta per partecipare a un TikTok dance trend. Nessun problema, zero esitazioni. Questo è un uomo che affronta con disinvoltura biodiversità, sedimentologia e, sì, persino una sfida virale ogni tanto, dimostrando un'apertura mentale senza limiti anche verso ciò che gli è completamente estraneo. Ed è proprio questa apertura che lo rende non solo rilevante, ma in continua evoluzione in un settore dove la tradizione spesso resiste al cambiamento.

Pensa a questo libro come a un regalo di compleanno reciproco: il suo dono per noi e il nostro per lui. Anche se, diciamocelo, è soprattutto un regalo per noi.

Ovviamente, il vero dono di Scienza non si esaurisce nella sua etica del lavoro implacabile (che è stancante persino da osservare), ma prosegue nella sua capacità di raccontare storie. Come ben sanno gli ascoltatori di Everybody Needs a Bit of Scienza, la serie di Italian Wine Podcast dedicata al Professore, Scienza può discorrere con passione su argomenti che spaziano da ampelografia e cambiamento climatico a mitologia e tettonica. Lancia intuizioni che potrebbero richiedere mesi per essere decifrate, con una naturalezza tale da dare l'impressione che lui ne stia appena scalfendo la superficie. E se può sembrare che parli a braccio, non farti ingannare: ogni risposta, ogni conferenza, ogni digressione inaspettata (e ce ne sono molte) è radicata in una vita intera di studio e in un'insaziabile curiosità. Non pretendo di comprendere sempre tutto quello che dice (del resto, chi potrebbe?), posso però dire con certezza che il suo entusiasmo è semplicemente contagioso.

Che dire poi della sua scelta di carriera? Scienza avrebbe potuto diventare un luminare in molte discipline. Nella sua infinita saggezza, però, ha scelto il vino... per nostra fortuna. Il suo contributo è stato così straordinario che spesso dico scherzando che, se parlasse inglese, sarebbe più famoso di Jancis Robinson. Ma, ahimè, non lo parla. Ed è proprio per questo che pubblichiamo questo libro in formato bilingue, per dare alle sue idee il pubblico più ampio che meritano. Se lo hai mai visto a Vinitaly, circondato da ammiratori e alle prese con un'agenda di appuntamenti che sembra umanamente ingestibile (che fa sembrare la mia quasi rilassante, e questo la dice lunga), allora sai esattamente quanto sia rispettato e richiesto.

Chi parla italiano, qui troverà Scienza nella sua forma più autentica e senza filtri, il più vicino possibile alla fonte originale (per così dire). Diciamo però la verità: Scienza non parla semplicemente italiano. Lui parla e scrive in "scienziese". Sì, tecnicamente è italiano, ma è l'italiano di Scienza. Tanto che, quando traduciamo i suoi libri, il processo avviene in due fasi: prima traduciamo dallo scienziese all'italiano, poi dall'italiano all'inglese. Perfino le sue sli-

de in PowerPoint richiedono una decodifica prima di poter essere condivise con il pubblico. Per fortuna, nel nostro team abbiamo quelli che chiamiamo affettuosamente i "coloro che sussurrano a Scienza", ossia esperti addestrati appositamente per sbrogliare le sue intricate costruzioni linguistiche. Per questo delicato processo hanno dato una mano Richard, Giulia e Michela.
Perché è proprio questo che rende la scrittura di Scienza così... "scientifica". È vivace, imprevedibile, a volte snervante. Le domande che pone sono più di quelle a cui risponde. Accende curiosità che non sapevamo di avere. Divaga, provoca e, proprio quando pensi di aver capito tutto, imbocca uno svincolo inaspettato per la filosofia antica o verso le crisi viticole del XIX secolo. Ed è proprio per questo che è così affascinante. Forse è il momento di ripassare i poemi omerici o di rispolverare gli appunti di biologia del liceo, ma è più di ogni altra cosa il momento di ripensare a come raccontiamo il vino, oltre i soliti cliché triti e ritriti.
Il titolo provvisorio di questo libro era Radici e ragione: l'evoluzione dello *storytelling* del vino, un titolo di cui eravamo piuttosto orgogliosi, un omaggio sia alle profonde radici storiche del vino, sia alla razionalità che ci aiuta a comprenderne il futuro. Inoltre, si allineava perfettamente con il nostro precedente libro di Scienza, Vine and Prejudice: Fake Science and the Search for the Perfect Grape. Alla fine, però, abbiamo optato per Lo *storytelling* del vino: un titolo forse meno letterario, ma più immediato per un pubblico italiano. Certo, manca quel tocco di stile presente nell'alternativa ispirata a Jane Austen. A ben vedere, tra l'altro, se da un lato Jane Austen potrebbe non aver letto Aristotele, i concetti aristotelici permeano la sua scrittura proprio come permeano il lavoro di Scienza.
Dalle pitture rupestri preistoriche ai viaggi mitologici di Ulisse, il vino è sempre stato un protagonista nella storia dell'umanità. Scienza riprende le antiche tecniche narrative - come il topos dell'e-

roe omerico o i concetti della retorica aristotelica - e le applica alla storia recente del vino, dalla lotta contro la fillossera alla nascita dei SuperTuscan. Ma, soprattutto, ci mostra come queste tecniche possano essere utilizzate oggi dalle aziende vinicole per raccontare le loro storie in modo che risuonino con il consumatore moderno. Ed è per questo che siamo ancora una volta entusiasti di condividere gli approfondimenti, l'esperienza e l'inarrestabile passione di questo straordinario narratore.
Buon compleanno, Professor Scienza! Che tu possa continuare a ispirarci, provocarci e, occasionalmente, anche esasperarci per ancora molti, moltissimi anni.

Professore ti vogliamo bene. Tanto..

Stevie Kim

Storytelling: la fabbrica di storie

Nel suo libro *Story or Die* (2021), Lisa Cron sostiene che gli esseri umani non siano fatti per leggere, attività per loro non naturale. Siamo invece nati per raccontare e ascoltare storie: le nostre menti si sintonizzano con i personaggi attraverso la voce del narratore, che risveglia le aree cerebrali deputate alle emozioni e ci permette di immedesimarci, di sentire empaticamente ciò che provano gli altri.

Lo storytelling, un'arte antica quasi quanto l'umanità, è sempre stato uno strumento cruciale per la condivisione di valori sociali. Ha radici profonde, che si possono rilevare già nelle prime forme di espressione artistica come i dipinti rupestri preistorici. Se le grandi narrazioni del passato raccontano miti universali che hanno plasmato la storia umana, lo *storytelling* moderno si concentra piuttosto sulla realtà contemporanea, costruendo cornici narrative che suscitano emozioni e offrono nuovi modi di interpretare il mondo. Dal diario di viaggio di Marco Polo alle avventure di Robert Louis Stevenson, il concetto di narrazione permea ogni ambito del sapere – dalla scienza politica all'antropologia, dalla ricerca medica alla biologia molecolare – arricchendo la nostra comprensione e dando vita a legami in virtù di esperienze condivise. L'avvento di Internet e delle tecnologie digitali, a partire dalla metà degli anni '90, ha segnato una svolta cruciale per lo storytelling, aprendo nuove frontiere e opportunità di connessione.

L'essenza dello storytelling, dove anche le storie moderne possono essere comprese attraverso la lente della leggenda, è ben descritta nelle parole della scrittrice israeliana Nomi Eve:

> «La parola "leggenda" deriva dal latino *legere*, che significa "leggere". La parola "finzione" deriva dal lati-

> no *fingere*, che significa "formare". Da *fingere* deriva anche la parola *fingers* ["dita" in inglese]. Formiamo cose con le dita. La parola "storia" deriva dal greco "*istor*", che significa "imparare" o "conoscere". Credo nell'etimologia originale. Credo che la finzione sia verità plasmata. Credo che la storia sia un modo di conoscere tutto ciò. Credo che la leggenda sia il modo in cui leggiamo tra le righe».

Le storie, sia antiche che moderne, sono strumenti potenti per comprendere il mondo e noi stessi. Questa capacità di plasmare la verità e di svelare significati nascosti è particolarmente evidente nella tradizione greca, dove il vino - elemento centrale di molte storie - diventa un mezzo di esperienza, un richiamo ai viaggi di Ulisse e dei coloni greci che portano la viticoltura in Occidente.

La forza di una storia

Avere una storia avvincente da raccontare è un elemento fondamentale e strategico, con un enorme potenziale per differenziare e rendere riconoscibile qualsiasi marchio. La domanda cruciale è: qual è l'essenza della storia che vogliamo narrare?

Ogni storia convincente scaturisce da un narratore ispirato. Per i produttori di vino, la missione deve trascendere il semplice profitto e comunicare un significato più profondo al proprio pubblico. Origini, aspirazioni, sfide e successi costituiscono le fondamenta di queste storie.

Definire la propria missione è il primo passo per costruire una narrazione risonante. Questo, però, richiede una tecnica precisa. Come distinguere una missione generica, formalmente corretta, da una che muove i cuori e ispira all'azione?

Creare una narrazione significativa implica l'approfondire il significato intrinseco della genesi e della produzione di un vino, l'impegno profuso e il motivo per cui dovrebbe essere scelto, persino desiderato. L'identità ultima di un vino risiede nella sua storia, non in un'etichetta accattivante.

Da qui inizia un viaggio, articolato in storie complesse che si intrecciano con la realtà del nostro mondo. Filosofia ed estetica non sono sufficienti per nobilitare un vino: è necessaria una comprensione profonda e autentica della sua origine e della sua produzione. Chiunque si sia occupato di scienza ha incontrato il tema della narrazione e della cultura greca. La narrazione è un processo radicato nella nostra memoria biologica e si basa su determinati *topoi*, schemi narrativi specifici e propulsori biografici che ruotano attorno alcuni elementi chiave: l'eroe in cerca di sé stesso, la ricerca, l'antagonista, il conflitto tra eroe e antagonista, il tesoro, la

scoperta che completa la ricerca, le avversità da superare, gli aiuti (umani o magici), il premio finale.
Ognuno di noi si sente un po' un eroe. Le nostre vite quotidiane riflettono il copione dell'eroe, con la ricerca di un tesoro alla fine del percorso. Omero, il padre della narrazione occidentale, sviluppa il tema dell'eroe che intraprende un viaggio alla ricerca di sé stesso, affronta esperienze avventurose e drammatiche, riceve aiuto e torna alla propria terra natale, tra i suoi cari.
In sintesi, la narrazione del vino e della relativa produzione deve catturare l'attenzione dei destinatari, creare nuovi significati, imprimere simboli e impressioni nell'immaginario collettivo, offrire una comprensione più profonda degli aspetti qualitativi di un vino e fornire contesto e rilevanza all'azienda che lo firma. Una narrazione efficace genera empatia e coinvolgimento, instillando fedeltà al marchio e al logo.

Il ruolo del consumatore

I consumatori pensano, ricordano e decidono sulla base delle storie. Come afferma Giulio Cesare nel *De bello gallico*,

> «Gli uomini credono volentieri ciò
> che desiderano sia vero».

Lo storytelling, vera e propria "fabbrica di storie", è un potente strumento di comunicazione: coinvolge attraverso la trama e la tensione narrativa, costruendo una fedeltà duratura. A differenza di un logo, che cattura l'attenzione in maniera immediata e veloce, lo *storytelling* agisce gradualmente e costruisce relazioni con il pubblico nel lungo termine. Il cervello umano elabora meglio le informazioni quando sono comunicate in forma narrativa: è per questo che mutua tecniche tipiche di romanzi, fumetti e cinema per trasformare schemi, figure e dati in racconti coinvolgenti.

Gli elementi immateriali di una narrazione sono molteplici: attrattiva del territorio, funzione di socializzazione e identificazione, proiezione di sé, adesione a uno stile di vita o di consumo, identità del produttore, mito o qualità. Un po' più semplici da interpretare sono gli elementi materiali: profilo gusto-olfattivo, funzione nutraceutica, salubrità, impatto sensoriale, fisiologico e psicologico, packaging, garanzia del marchio, prezzo. Tutti questi elementi insieme costituiscono da un lato la sostanza fisica di una bottiglia, dall'altro la sua anima. È difficile costruire una storia efficace basandosi unicamente su elementi materiali.

La storia del vino inizia quindi da quella del territorio, con la relativa storia, l'antropologia, la geografia, il suolo e il clima. Questi elementi sono fondamentali per il racconto perché concorrono a evocare il senso autentico di un luogo. Non possiamo diffondere efficacemente la storia di un vino senza ancorarlo a una parvenza di realtà. Il linguaggio più efficace è quello dei sensi: il territorio può essere descritto non solo fisicamente e didascalicamente, ma anche descrivendo l'interazione delle sue caratteristiche coi nostri sensi e le reazioni risultanti.

Naturalmente, non possiamo prescindere dal vino stesso: se non abbiamo un prodotto di qualità da raccontare, non ha alcun senso parlare del suo territorio. Anche il profilo del produttore, lo stile di consumo del vino e la genesi del mito sono molto importanti. Proprio sul mito si può costruire la storia dell'eroe. Nessun'altra categoria di prodotti ha sviluppato una poetica tanto raffinata, complessa e dettagliata come quella della produzione del vino. L'enologia si trasforma così in "enosofia", e il vino può elevarci da una razionalità troppo umana allo *status* di simbolo sacro. È tuttavia fondamentale che imprenditori, comunicatori del vino e addetti alla vendita al dettaglio non cadano nella trappola della retorica e del ricco repertorio di fiabe che hanno contribuito a creare e sviluppare alcune riflessioni non conformiste sulla possibilità di adattare il vino ai gusti dei consumatori, soprattutto quelli di domani.

C'è bisogno di una storia

Il vino nasce dall'incontro tra vite e tecniche di vinificazione dalle origini lontane. Solo con il progresso della tradizione si è giunti a una sintesi perfetta, sperimentabile nei vini che conosciamo oggi. Nessun altro prodotto agroalimentare riesce a soddisfare quanto il vino il desiderio di eclettismo e sincretismo del consumatore moderno. Eclettico è chi si muove con disinvoltura tra stili diversi; sincretico è chi riesce a fonderli in modo armonioso.
Oggi è sempre più difficile distinguere l'influenza del gusto estesico, legato alle proprietà sensoriali di un vino o di un cibo, da quella del gusto estetico, la capacità di discernere bellezza e imperfezione. Nel caso del vino è fondamentale distinguere ciò che è differente da ciò che è indifferenziato, un processo molto gratificante per gli appassionati perché capace di generare una sorta di euforia in un contesto generalmente disforico, che tende all'omologazione e all'appiattimento. Questo spinge il gusto collettivo verso una crescente richiesta di varietà nelle tipologie di vino a disposizione, le quali sono spesso distinguibili solo dall'etichetta.
In passato, il concetto di "buon gusto" era determinato o dall'egemonia di una classe sociale, ad esempio la borghesia del tardo Settecento, o dal "pensiero unico" di alcuni critici, capaci di catalogare i vini secondo profili sensoriali ben definiti; si pensi ai Bordeaux. La recente storia del consumo di vino mostra però una serie crescente di anomalie che includono o escludono determinati vini o addirittura interi territori dai circuiti elitari di pochi brand dominanti.
Ormai tutti percepiamo non solo il significato della globalizzazione, ma soprattutto il suo impatto. Questa non riguarda solo i grandi eventi globali, ma influisce anche sulle nostre abitudini quotidiane, compreso il nostro rapporto con il vino. La globaliz-

zazione ha portato vantaggi evidenti, specialmente dal punto di vista commerciale, ampliando la diffusione del vino in culture che ne ignoravano l'esistenza. Tuttavia, non vanno sottovalutati gli svantaggi. La crescente standardizzazione dei vini, spesso indistinguibili tra loro indipendentemente dal vitigno o dal terroir, è il risultato di un "pensiero unico" che impone modelli sensoriali precisi di vini perfetti in relazione a determinati canoni, ma privi di anima.

A questa tendenza si è risposto con forme alternative di vinificazione e viticoltura: il recupero di vitigni autoctoni, la fermentazione in anfore, la viticoltura biodinamica. Tuttavia, molti vini ottenuti da vitigni autoctoni, caratterizzati da profili sensoriali originali, faticano a essere accettati nei modelli descrittivi adottati da chi assegna punteggi e valutazioni per le guide di settore.

Il problema è di indubbia importanza per il marketing: chi produce, comunica e promuove il vino deve riuscire a metterne in luce gli aspetti più rilevanti. Spesso, il consumatore è attratto maggiormente dalle sfumature linguistiche con cui il vino viene descritto che dal giudizio tecnico degli esperti, i quali non sempre riescono a influenzarlo in modo significativo. Oggi sono soprattutto il fascino dei luoghi di produzione, l'attrattiva de territori e l'accoglienza delle cantine a trasmettere la carica evocativa del vino all'immaginario dell'enonauta. Tuttavia, questa esperienza rimane un privilegio di e per pochi.

Il destino del vino, che piaccia o no, è nelle mani della grande distribuzione, che investe in modo massiccio nel settore e punta su prezzi competitivi, un ampio assortimento di prodotti e un'atmosfera evocativa nelle enoteche. Questo, che magari non sorprende gli addetti ai lavori, evidenzia un deficit di strategie di marketing capaci di reagire ai cambiamenti di una società sempre più fluida, dove circolano troppe informazioni e il consumatore fatica a estrarre ciò che ha un reale valore per sé stesso. Assistiamo così ad un crescente processo di de-semantizzazione del linguaggio e di logoramento dei processi simbolici.

La speranza è che possa esserci, tra produttore e consumatore, un rinnovamento nei rapporti capace di trasformare le minacce in opportunità. In estrema sintesi, ciascun produttore deve trovare una risposta convincente da fornire al consumatore che si chiede: «Perché dovrei scegliere di bere proprio questo vino?». L'espressione oraziana *miscere utile dulci*, letteralmente mescolare l'utile con il dilettevole, è perfetta in questo contesto. Nell'*Ars Poetica*, Orazio raccomanda di usare la narrazione per ispirare il lettore, combinando sapientemente un po' di divertimento e un po' di insegnamento.

Il mondo cambia, la società si evolve e il vino non è immune ai fenomeni che influenzano l'intero universo dei consumi. Anzi, per il suo valore simbolico e sociale, il vino è uno dei settori più esposti a questi cambiamenti. Stiamo assistendo a una rivalutazione del vino come aggregatore piuttosto che come distintivo di *status*, con ripercussioni importanti sul mercato. Il consumatore non fonda più le proprie scelte sulle pulsioni dell'ego nel tentativo di distinguersi e affermare la propria individualità, ma si lascia guidare da fenomeni di emulazione e condivisione. Oggi, bere vino significa riscoprire le proprie radici, tornare in luoghi dimenticati, celebrare un ricordo, rinnovare un senso di fraternità o rivivere un'esperienza comunitaria.

Il verbo "eccellere", dal latino *excellĕre* (spingersi fuori), non indica solo un livello superiore all'interno dello stesso ambito, ma anche il passaggio a una dimensione esterna, diversa. Per trasformare un vino da "buono" a "eccellente", dunque, non basta spiegarne le caratteristiche intrinseche: serve "spingere fuori" l'attenzione e comunicarne le intenzioni e l'impegno dei produttori, i valori, la

visione. Non si tratta solo di presentare il prodotto, ma anche di coinvolgerne il produttore, la relativa ispirazione e il metodo che ha seguito.
Se basata solo sulle caratteristiche sensoriali, la qualità di un vino può essere sufficiente o buona. Per essere eccellente, però, deve contemplare anche valori immateriali che non risiedono solo nel vino stesso, bensì nel suo spirito, nell'umanità di chi lo ha prodotto e nella relazione con chi lo degusta; il primo elemento immateriale che il consumatore percepisce è la credibilità del produttore.

Emozioni, simboli e comunicazione del vino

Siamo preda delle nostre emozioni, ma se così non fosse non potremmo partecipare in modo significativo agli eventi della nostra vita e relazionarci con quella degli altri. Saremmo incapaci di apprezzare le informazioni che ci arrivano dall'arte, dalla società e dalla scienza. Le emozioni sono inoltre collegate ai nostri sensi, ossia vista, udito, tatto e gusto: sarebbe possibile apprezzare un dipinto, un brano musicale o il sapore di un buon vino senza di esse? Potremmo chiederci se le emozioni siano l'espressione di puri e semplici meccanismi di adattamento comportamentale o, piuttosto, il risultato di "costruzioni", ovvero una sintesi che integra e amalgama input sensoriali, tracce mnestiche, micro-comportamenti, routine testuali e aspettative. L'emozione non è solo fondata su percezioni fisico-psichiche, ma esige connessioni con l'immaginazione, intesa come risultato della compatibilità tra la mente e il mondo esterno, il cosiddetto senso comune.
Nel 1995, Daniel Goleman scrive un libro intitolato *L'intelligenza emotiva* nel quale la descrive come uno strumento che ci consente di leggere le emozioni, di tessere relazioni, di conoscerci più a fondo e valutare le reazioni di chi ci sta di fronte. La sede delle emozioni è nel nostro cervello, in ciascuno dei lobi temporali, in una struttura a forma di lacrima chiamata amigdala che interpreta gli stimoli provenienti dal mondo esterno: l'angoscia, la curiosità, il desiderio, l'empatia, la nostalgia, la sorpresa nascono qui.
Le emozioni possono anche plasmare le nostre percezioni. I vini dell'Etna, ad esempio, sono celebrati per la loro qualità: quanto questa opinione è però influenzata dalla meravigliata constatazione che provengono da suoli di origine vulcanica? Quanto la nostra percezione della bellezza di Bolgheri è accresciuta dal

fatto che, sul suo territorio, vi si produce il Sassicaia? L'emozione del gusto è prodotta dall'esercizio dell'immaginazione.
Nel codice manageriale dell'*experience economy*, la produzione di plusvalore è strettamente legata alle cosiddette "esperienze emozionali" e alla sfera dell'immaginario attraverso la produzione di artefatti ad alta componente simbolica. L'empatia è la capacità di riconoscere ciò che un'altra persona sta pensando o provando e di reagire con uno stato emotivo simile, provando dunque un'emozione condivisa. Oggetto delle ricerche di Giacomo Rizzolatti, neurofisiologo e Professore emerito dell'Università di Parma, i neuroni specchio sono una classe particolare di cellule nervose presenti in diverse zone della corteccia motoria del cervello. Vengono attivati quando un essere umano compie un'azione e, contemporaneamente, questa azione si verifica anche nell'osservatore dell'azione stessa: per questo, sono importanti nell'apprendimento attraverso l'imitazione. Possono essere definiti anche i neuroni dell'empatia o della partecipazione empatica e hanno perciò un ruolo significativo nelle relazioni interpersonali.
L'educazione al sorriso e alla gentilezza al fine di trasmettere questo atteggiamento agli altri dimostra che il sorriso viene trasferito attraverso i neuroni specchio, quasi come se disponessimo di un secondo cervello – parallelo a quello preposto al pensiero razionale – dedicato alle relazioni sociali, alle emozioni e alle azioni. Secondo alcune teorie, l'autismo sarebbe correlato con disfunzioni di tale sistema. Lo stesso meccanismo si manifesta nel teatro e nello sport: un giocatore di basket, ad esempio, può intuire dalle mani dell'avversario se il tiro andrà a segno; nelle manifestazioni collettive, specialmente in quelle che evocano sentimenti profondi, i neuroni specchio si attivano e lo spettatore partecipa come se fosse presente nella scena.
La vera innovazione portata dal *cinématographe* dei fratelli Lu-

mière è stata quella di trasformato il cinema da un'esperienza individuale a un evento condiviso da un pubblico numeroso. Similmente, il vino è diventato un fenomeno sociale, passando da semplice esperienza gastronomica a esperienza emotiva condivisa. La psicologia del consumo ha dimostrato l'importanza del pubblico come elemento immaginario nella mente del produttore di vino, un pubblico che può essere costituito anche un singolo individuo che si lascia trasportare dalle suggestioni e dagli stimoli sensoriali del vino.

Qual è, dunque, il contributo dei neuroni specchio nella creazione di un clima di empatia tra i partecipanti a una degustazione e, in ultima analisi, nella percezione positiva delle caratteristiche sensoriali di un vino? Nei comportamenti collettivi delle degustazioni moderne possiamo riconoscere l'antico rituale simbolico del simposio greco e della messa cristiana, definibile come "principio di incorporazione". Il consumatore odierno si confronta inconsciamente con due problemi sostanzialmente nuovi: la distruzione del cibo e del vino nella loro forma materiale da un lato, la loro assimilazione nei tessuti del proprio corpo dall'altro. Sono in fondo due aspetti insiti nell'atto del nutrirsi che il mondo classico risolve attraverso i riti dionisiaci: il dio muore due volte, con la pigiatura dell'uva e, dopo la rinascita attraverso la fermentazione, durante l'atto del bere il vino. Analogamente, nella mistica cristiana i due momenti sono la morte – con la successiva resurrezione – di Cristo e l'eucaristia.

Analizzando la fenomenologia della messa, Carl Jung riconosce nell'atto di "mangiare il dio" il significato del termine greco *syn-ballein* (da cui deriva "simbolo"), ossia "essere insieme", l'alchimia che trasforma la materia grezza del vino in una nuova realtà psichica, la sottomissione della realtà biologica al bisogno spirituale che rafforza il legame sociale.

In passato, il vino era al centro di diverse manifestazioni delle società occidentali come festeggiamenti di nascite, celebrazioni di matrimoni, riti di passaggio fino alla sepoltura dei defunti. Oggi è in atto un fenomeno di de-simbolizzazione che testimonia il progressivo allontanamento dell'uomo moderno dalle suggestioni del mito. Quali sono le conseguenze di questo sulla sociologia del consumo del vino? Tanto nel simposio greco quanto nell'odierno consumo condiviso attorno a un tavolo, magari accompagnato da conversazioni interessanti, l'aura evocativa del vino difficilmente sfocia nell'ebbrezza. In altri contesti, invece, il consumo smodato di alcol punta quasi dichiaratamente al fine di raggiungere l'ubriachezza; in questi casi, bere a stomaco vuoto è ben meno grave del bere a cervello vuoto.

Volenti o nolenti, dobbiamo tornare al significato religioso del vino, inteso come valore profondo non tanto legato alla fede in sé, quanto più al senso di appartenenza che unisce una comunità tramite usi e tradizioni. Le comunità sono di norma legate a un territorio: il vino, i cui profumi evocano il paesaggio e la terra madre, è testimone di questa connessione. Chi rifiuta il vino, in un certo senso, nega con forza la propria appartenenza a quella terra.

Negli ultimi anni abbiamo assistito a una proliferazione di vini, gestita principalmente da esperti di marketing e nella quale gli uomini della vigna sono relegati sullo sfondo, con un ruolo minimo. In questo trionfo dell'immagine sulla sostanza, si è cercato di creare un'aspettativa quasi magica di piacere e godimento, senza badare a spese per effetti speciali. Il momento della verità, quello cioè del confronto tra il tanto celebrato vino e le papille gustative del consumatore, ha però spesso rivelato una discreta discrepanza tra quell'aspettativa e la realtà, sia dal punto di vista sensoriale che emotivo. Questa sequenza di adescamento, illusione e conseguente delusione mina la credibilità dei produttori e si traduce in

una progressiva sfiducia del consumatore. Come l'amante sedotto e abbandonato, quest'ultimo finisce dunque per allontanarsi dal vino e gettarsi fra le braccia di altre categorie di bevande.
La comunicazione del vino deve muoversi verso una dimensione più sobria e trasparente, che crei canali diretti tra produttori e consumatori, che valorizzi la relazione del prodotto con il contesto produttivo e di consumo. Dobbiamo cercare una comunicazione che evochi nel destinatario significati e non si limiti a trasmettere informazioni tecniche: questo può avvenire raccontando il vino attraverso la passione e l'impegno del produttore, mettendo in luce i valori e i legami che uniscono chi lo produce e chi lo sceglie.
In definitiva, abbiamo bisogno di una comunicazione che si abbandoni al piacere del raccontare e di narrare storie.

Comunicare il vino: un'arte per pochi

È in corso un grande cambiamento nell'approccio alla valutazione del vino. Come accade nel cinema o nell'editoria, dove la critica può decretare il destino di un film o di un libro, anche nel mondo del vino il giudizio può rappresentare un punto di svolta. Un tempo, i commenti dei critici offrivano prospettive interessanti, aprivano nuovi orizzonti e introducevano idee sorprendenti. C'era poi il giudizio della gente comune, di chi beveva un vino, leggeva un libro e ne parlava con gli amici.

Con l'avvento dei social media, però, tutto è cambiato. In primo luogo, perché la democrazia del web mal sopporta l'autorevolezza del critico (il quale, a dirla tutta, non sempre è realmente autorevole). In secondo luogo, perché il buon senso non può ridursi a un semplice "mi piace" o "non mi piace". Oggi leggiamo opinioni, giudizi ed espressioni riprese da ogni dove; cerchiamo di dimostrare di saperne più degli altri, ma spesso ci mancano l'esperienza e la conoscenza necessarie. Un risultato, tuttavia, è stato raggiunto: è stato smantellato il pensiero unico imposto dagli *opinion leader* che per anni hanno dettato i canoni della qualità non sulla base del reale valore dei vini, bensì su quella dei propri interessi.

Quando si presenta un vino, è fondamentale evitare un linguaggio ridondante e artificioso, pieno di figure retoriche e frasi altisonanti. Abbandoniamo la falsa distinzione tra una valutazione tecnica e un giudizio basato sul piacere puro; smettiamo di trasformare le schede tecniche in strumenti onnipresenti, che pretendono di descrivere, analizzare, classificare e persino celebrare il vino, finendo per renderlo impersonale.

Le storie vivono di emozioni; la nostra stessa vita è fatta di emozioni. Le narrazioni più coinvolgenti sono quelle che mettono in luce la vulnerabilità, permettendoci di identificarci con il protagonista e suscitando una risposta emotiva intensa. Una storia cattura la nostra attenzione, ci trascina dentro il proprio mondo e sostituisce temporaneamente la nostra realtà con la propria. Aristotele insegna che una storia ben costruita, con un inizio, uno sviluppo e una conclusione, si regge sulla struttura della trama. Essa è ciò che vediamo e che ci affascina, ma non è tutto: ciò che la rende davvero coinvolgente è il percorso interiore del protagonista che si svolge al suo interno, il processo che lo porta a una nuova consapevolezza grazie alla quale supererà l'ostacolo che gli impedisce di risolvere il problema.

Per Nietzsche non esistono fatti, ma solo interpretazioni. Gli esperti di comunicazione hanno scoperto che il cervello umano assimila meglio le informazioni quando queste vengono presentate in forma narrativa, coinvolgendo più sensi contemporaneamente. Lo *storytelling* è uno strumento potente per catturare l'attenzione, a patto però che il linguaggio si sviluppi attraverso un discorso di verità fondato su precise condizioni. Nel proprio seminario sulla *parrhesia* (dire la verità, parlare con franchezza) e riferendosi alla crisi della democrazia ateniese tra il IV e il V secolo a.C., Michel Foucault elabora a questo proposito il concetto del "rettangolo", un modello che permette di sincronizzare idealmente quattro condizioni: la forma (competizione leale), la realtà dei fatti (il potere dei monopoli), la verità (un *logos* ragionevole) e l'etica (il coraggio della lotta). Il tema è molto attuale se considerato in relazione al rapporto tra social media e propaganda di ogni genere, non solo politica. Nati come strumenti di libertà e democratizzazione dell'informazione, talvolta anche potenziati dall'intelligenza artificiale, i social sembrano essersi trasformati in un inquietante mezzo di disinformazione e manipolazione di massa, con la diffusione ormai incontrollata delle *fake news*.

Vino e società: un rapporto in evoluzione

Le ricerche di mercato rivelano che, tra gli italiani che bevono vino regolarmente (circa il 60% della popolazione), solo il 2% si considera un intenditore, mentre il 20% ritiene di avere una competenza accettabile. Emerge inoltre una tendenza che vede il vino abbinato al cibo in modi sempre meno convenzionali. Il vino ideale, soprattutto per i consumatori più giovani e per quelli dei nuovi mercati, sembra discostarsi nettamente dall'offerta attuale, modellata dall'estro dei produttori e dalla tradizione.
Le donne preferiscono di norma vini speziati, aromatici e floreali, mostrando maggiore apertura alle novità (compresa la cucina etnica). I vini tannici e strutturati restano appannaggio soprattutto degli uomini, tra i quali cresce però l'interesse per vini più leggeri e meno alcolici. Si delinea quindi una progressiva distanza tra le esperienze e le aspettative dei consumatori e l'approccio spesso rigido dei produttori, fenomeno che evidenzia la necessità di rivedere il modo in cui il vino viene comunicato oggi.
Storicamente radicato nella cultura mediterranea, il vino ha sempre rispecchiato i cambiamenti della società. Per secoli, il vino di bassa qualità è stato un alimento di base per le classi più povere. Fino agli anni '60 e '70, era prevalentemente "fatto in casa", venduto sfuso e acquistato localmente. Con l'allontanamento della società italiana dalle sue radici agricole, il consumo di vino ha iniziato a seguire la logica dei beni di lusso: le masse hanno iniziato a emulare le abitudini delle classi privilegiate, appropriandosi di un linguaggio culturale, costruendo cliché e, soprattutto, trasformando il vino in un argomento di conversazione.
L'abitudine a bere alcolici non si sviluppa spontaneamente, ma si acquisisce per imitazione di comportamenti familiari e sociali. Oggi, i giovani bevono raramente vino in casa e frequentano

ambienti in cui il vino gioca un ruolo quasi nullo. Dove possono allora apprendere quel vocabolario sensoriale indispensabile per comprenderlo? Per le nuove generazioni, il vino soffre di una mancanza di riferimenti simbolici, nonostante i numerosi tentativi di colmare tale lacuna attraverso iniziative educative. La cultura del vino è diventata un sapere libresco calato dall'alto, veicolato dai media e dominato da esperti che, però, il più delle volte non sono abbastanza carismatici da convincere il pubblico.

Come abbiamo già ricordato, alla centralità del vino in diversi momenti chiave per le società occidentali del passato si contrappone oggi un processo di de-simbolizzazione, specchio della progressiva perdita del fascino mitico nella società contemporanea. Il mondo del vino è intriso di simboli: di alcuni siamo consapevoli, di altri invece percepiamo l'esistenza nel nostro inconscio, ma non riusciamo a metterne a fuoco i contorni. A molti dei nostri riti si possono collegare miti legati al vino. Affinché un rituale mantenga la propria efficacia, però, è necessario che i partecipanti ne condividano lo stesso mito e trovando in esso un codice comune di appartenenza. Quando il vino esce dalle abitudini di un gruppo specifico e fa il proprio ingresso in un contesto globale, il suo linguaggio simbolico perde il significato originario e si trasforma in nuove espressioni rituali e forme di comunicazione differenti. Ad esempio: siamo certi che il linguaggio con cui comunichiamo il vino nei paesi occidentali – fortemente influenzato dalla mitologia greca – sia altrettanto efficace per i popoli dell'Estremo Oriente, le cui tradizioni culturali si basano su miti e simboli molto diversi dai nostri? Qual è il rapporto tra la presunta universalità del mito e la tradizione?

La risposta sta nella natura stessa del linguaggio simbolico, che non appartiene a un sapere universale facilmente condivisibile tra tutte le civiltà, ma è piuttosto l'espressione di conoscenze radicate in contesti locali, trasmesse attraverso quell'arbitrarietà mitologica che gli antropologi identificano nella divinità fondatrice, il principio della storia. E cosa custodisce meglio il senso della storia, se non la tradizione?

Il rischio, però, è che il rito si riduca a una formalità priva di contenuto, svuotata del proprio significato nel corso delle continue reiterazioni. La soluzione sta nell'arricchimento costante attraverso elementi di "ampliamento rituale" quali una maggiore consapevolezza tecnico-scientifica sui processi che determinano i profili sensoriali del vino (come l'origine geologica del suolo, le caratteristiche climatiche, la varietà delle uve) o sulle fasi della produzione. Ricerca e conoscenza sono fondamentali per far crescere la cultura del vino, così come l'attenzione per la materia prima, in un'ottica olistica che tenga conto della biodiversità, della conservazione delle tradizioni secolari e degli aspetti antropologici e sociologici legati alla produzione vitivinicola.

La nuova era narrativa

A partire dalla fine del XX secolo, lo *storytelling* dei brand negli Stati Uniti e in Europa si trasforma sotto i meccanismi dell'industria dei media e dal capitalismo globalizzato, diventando uno strumento potente per influenzare le opinioni dei consumatori e dei cittadini. Un tempo riservato ai bambini, attraverso cartoni animati e racconti delle nonne, lo *storytelling* si è evoluto in quello che è stato definito il *narrative turn*, ovvero una nuova era della narrazione. Parallelamente, nasce la narratologia, la disciplina che studia il raccontare storie. Le strategie di marketing si sono spostate da un approccio centrato sul prodotto a uno focalizzato prima sul logo, poi sulla narrazione. Oggi, il successo di un brand dipende molto dalla storia che lo racconta.

La crisi dei brand nasce dal comportamento sfuggente, imprevedibile e a tratti paradossale dei consumatori, che non sono più affascinati dal semplice marchio. I brand hanno iniziato a parlare e, quando lo fanno, i consumatori ascoltano con attenzione. «*Cosa c'è in un nome?*», si chiedeva Shakespeare in *Romeo e Giulietta*. Nel 1996, le immagini di bambini pakistani curvi su palloni da calcio con il logo Nike fecero il giro del mondo, rivelando storie di miseria e sfruttamento celate all'interno del brand. Levi Strauss ha costruito la propria identità attorno all'archetipo dell'eroe della classe operaia, vendendo jeans attraverso il racconto della vita dei minatori. Questo universo narrativo, una vera e propria "dream society", apre le porte a una fantasia condivisa e connette le persone su scala globale. Il consumo diventa così una sorta di comunione planetaria che ricorda la comunione dei santi cattolici. Consumando simboli mitici, come il vino, interiorizziamo il mito più potente di tutti i tempi: l'eroe. *Guerre Stellari* ne è un eloquente esempio contemporaneo.

Lo *storytelling* è la scienza che trasforma e promuove elementi – reali o immaginari – in parole, immagini, suoni e percezioni concrete, rendendoli così autentici e legittimati a esistere. È un approccio comune a molte discipline, dall'antropologia alla genetica, dalle neuroscienze alla geologia. Si è affermato anche come strumento manageriale imprescindibile per farsi ascoltare e scegliere.
La cultura pop è ossessionata dallo storytelling. La settimana della moda maschile, che si tiene ogni anno a Milano in gennaio e poi a Parigi in giugno, è un tripudio di narrazioni. Più *pathos* c'è, più la storia funziona. Le sfilate hanno ormai raggiunto livelli di teatralità mai visti prima: gli abiti sono solo uno degli elementi in scena, talvolta nemmeno il principale, in un allestimento complesso che punta a stimolare i sensi.
In generale, nel nuovo *storytelling* prevalgono l'ambiente urbano tutte le relative criticità. Oggi, ciò che conta è soprattutto l'immagine. Per comprendere il valore estetico e il percorso creativo delle collezioni è necessario essere consapevoli della rappresentazione che si ha davanti: gli abiti vanno visti come maschere, come costumi, non come semplici capi d'abbigliamento. Il pericolo latente di una disconnessione tra ciò che appare in passerella e la realtà effettiva della vendita si è ormai concretizzato. La finzione prende il sopravvento. Gli abiti delle sfilate sono semplicemente un'esca per i social network: nei negozi, troveremo tutt'altro. Questa divergenza è sempre esistita: passerelle e riviste hanno bisogno di trovate sceniche ad alto impatto, mentre i vestiti destinati alla vita di tutti i giorni devono anche rispondere a criteri di funzionalità. E se questa stessa dicotomia fosse arrivata anche nel mondo del vino?

Dall'*Odissea* al digital *storytelling*

La crescente sfiducia nella pubblicità tradizionale, la cosiddetta "comunicazione di brand", è una diretta conseguenza dell'avvento di Internet, che ha innescato un cambiamento profondo. Insieme ai social media, ha dato origine al *digital storytelling*, segnando il passaggio da una comunicazione asimmetrica, attuata dai brand con ingenti investimenti pubblicitari e messaggi ripetuti sui media tradizionali, a una condivisione narrativa. In questo nuovo scenario, la storia suggerita dal brand viene integrata e reinventata dal pubblico grazie alla creazione di spazi di dialogo dove ciascuno può esprimersi liberamente e condividere esperienze. I blog svolgono egregiamente questa funzione.

I contenuti pensati per comunicare un'idea si articolano in post, podcast, foto, video, commenti su YouTube. Nascono e viaggiano sul web, dove si arricchiscono di dettagli inaspettati e si trasformano in narrazioni interattive, coinvolgendo una moltitudine di utenti. Questo *storytelling* sfrutta gli strumenti digitali per creare storie multimediali dall'alto impatto emotivo. Le storie digitali sfruttano al massimo il potenziale espressivo di immagini, filmati, musica e racconti di vita vissuta. Tale approccio è utilizzato principalmente in ambito didattico, formativo e aziendale, sia per trasmettere i valori di un brand che come strumento di marketing o comunicazione interna. I casi di studio sono numerosi, anche se mancano ancora formatori ed esperti di *storytelling* soprattutto a causa di uno scarso sviluppo nella cultura dei linguaggi audiovisivi e multimediali.

Oggi, grazie alla tecnologia digitale e a YouTube, chiunque possieda una videocamera e abbia una storia da raccontare può trovare il proprio pubblico. Il fenomeno della condivisione di video online è destinato a diventare un canale di comunicazione sempre più

rilevante. I blog, come esempio di narrazione, rappresentano una forma di *storytelling* capace di superare le barriere della comunicazione aziendale impersonale, permettendo un'interazione più diretta e innovativa con i diversi interlocutori, interni ed esterni.
L'uso dello *storytelling* in differenti contesti narrativi è uno degli strumenti più efficaci per catturare l'attenzione di un pubblico sempre più distratto da infinite connessioni digitali. Ma questo approccio è davvero una novità? Il successo dell'*Odissea* di Omero è legato in gran parte alla sua capacità di raccontare i fatti secondo una trama funzionale alla conquista dell'Occidente da parte degli Eubei navigatori. Possiamo considerare quest'opera come un manuale di viaggio a tutti gli effetti, pensato per esorcizzare le paure degli equipaggi in rotta verso l'ignoto. Non è un caso che i luoghi in cui sono stati ritrovati *skyphoi* euboici (vasi con semicerchi disegnati lungo il bordo utilizzati dagli antichi greci come coppe da vino) coincidano con le tappe del viaggio di Ulisse descritte da Omero.
La nostra percezione della storia dei marinai greci fatica a distinguere la verità dalla finzione, la realtà dal mito. Il punto di forza della letteratura greca sta proprio nella sua relazione con la realtà, con lo spazio, con l'idea di frontiera e conquista, attraverso il costante riferimento a luoghi storici e ai miti del Mediterraneo. Per lungo tempo, il Mediterraneo e le sue coste sono stati più di una semplice meta celebrata nei racconti: rappresentano un vero e proprio orizzonte narrativo verso cui si dirigevano i popoli del vicino Oriente, un luogo dove tutto sembrava possibile e dove chiunque poteva iniziare una nuova vita. Il Mediterraneo era, allo stesso tempo, nazione e narrazione.

Gli elementi strutturali dello *storytelling*

Le principali parti costitutive di una narrazione sono la storia, che si concentra sul "cosa", e il discorso, che si concentra sul "come". Nella storia, i personaggi si muovono all'interno di un ambiente definito e interagiscono attraverso azioni ed eventi. Il discorso, invece, si articola sulle strutture scelte per la trasmissione e sulla manifestazione nei media.

Di norma, la narrazione si sviluppa secondo due modelli: il modello onnicomprensivo, che si sviluppa in saghe - epopee e leggende - e storie, come racconti brevi, podcast e romanzi, e il modello frammentario, usato principalmente dai media sotto forma di gossip, notizie e indiscrezioni. Si tratta di due modalità comunicative differenti: nel secondo caso, lo *storytelling* è rapido e non esaustivo, pensato per suscitare curiosità; nei racconti brevi o nei romanzi, invece, la narrazione offre spunti e temi che possono essere sviluppati anche attraverso altre forme artistiche, come il cinema o il teatro.

Una narrazione efficace nel mondo del vino deve combinare quattro elementi chiave: il racconto e la messa in scena, l'effetto subliminale del vocabolario e delle metafore, lo stimolo e la tensione narrativa - elementi fondamentali in un romanzo per coinvolgere il lettore e spingerlo a continuare la lettura - e l'effetto contagio, ovvero la diffusione virale di una storia che assume un ruolo sempre più rilevante su internet e social network.

Lo *storytelling* del vino necessita un approccio interdisciplinare che coinvolga tutti gli elementi alla base della sua produzione. Le macroaree narrative comprendono sia aspetti umanistici - come i contenuti storici, letterari, archeologici, antropologici - sia competenze scientifiche, ad esempio quelle relative al terroir e alle varietà di uva. Tale interdisciplinarità rappresenta una sfida, poiché è raro trovare persone che possiedano l'intero range delle competenze necessarie e siano allo stesso tempo in grado di raccontare il vino in modo efficace attraverso lo storytelling.

Come in ogni storia ben costruita, anche nello *storytelling* è fondamentale la presenza di un narratore. Ma qual è il suo ruolo? Due sono gli aspetti essenziali: la declinazione della storia in un genere letterario (thriller, mistero, romantico, saggistica scientifica, fantasy) e la scelta della trama (epica, tragedia, melodramma, commedia). Il narratore deve inoltre contestualizzare il tempo della storia e armonizzarlo con lo sviluppo degli eventi.

Venezia, Malvasia e nascita dello *storytelling* moderno

Venezia, con il commercio delle Malvasie, ci offre un esempio paradigmatico di *storytelling* che, grazie a una strategia di comunicazione anticipatoria, esce dai canoni della narrazione classica fino ad allora utilizzata nel commercio vinicolo del Mediterraneo e si basa su uno schema molto semplice e due strumenti ancora attuali: il brand (il nome Malvasia) e la notorietà (un vino esclusivo, per le corti europee).

Venezia intuisce che il vino deve essere trasformato da semplice genere alimentare a oggetto di culto, a icona. Se i Greci avevano associato il vino a Dioniso per l'uso rituale nel simposio, Venezia carica i vini del Mediterraneo orientale di valori laici, trasformandoli in una moda, in un rimedio per il corpo e lo spirito in un'Europa afflitta da carestie e dalla Peste Nera, conseguenze della cosiddetta Piccola Glaciazione iniziata nel XIV secolo.

La Serenissima non si serve né del simbolismo greco di morte e resurrezione (dall'uva al vino), né del significato mistico del sangue nel rito cristiano. Punta invece su un concetto più moderno: il vino come *status symbol*, un'idea vicina ai modelli dell'antica Roma imperiale, basata sull'origine e sulla rarità del prodotto. Adottando *ante litteram* alcune delle più efficaci strategie di comunicazione contemporanee, Venezia lega quei vini a un consumo elitario, donandoli ai sovrani dell'epoca e servendoli nei banchetti ufficiali alla presenza di diplomatici provenienti da tutta Europa, imponendo inoltre dazi doganali elevati per limitarne l'accesso alle sole classi abbienti. Allo stesso tempo, valorizza il senso dell'origine e della provenienza da territori lontani avvolti nel mistero come quelli sotto il controllo bizantino.

Un aspetto rivoluzionario è l'utilizzo del nome del luogo di produzione per commercializzare il vino, novità assoluta in un'epoca in cui i vini erano designati con indicazioni molto generiche quali *de plano, de monte* o semplicemente con il nome del vitigno. Così, Venezia offre ai consumatori un vino dolce e aromatico, fino ad allora sconosciuto in Occidente.

Sinestesia e sindrome di Proust

La sinestesia, in senso stretto, è un fenomeno neurologico in cui una stessa stimolazione genera automaticamente due risposte sensoriali distinte e concorrenti. Di norma, nella percezione sensoriale - sia essa visiva, uditiva, olfattiva o tattile - una modalità prevale sulle altre. Nella condizione sinestetica, al contrario, vedere una lettera o un numero porta spontaneamente a riconoscervi un determinato colore; ascoltare una determinata musica o il suono di una parola fa contemporaneamente percepire un sapore specifico. Non si tratta di un'associazione di idee, bensì una vera e propria esperienza sensoriale.

Per estensione, si prende in prestito il concetto di sinestesia anche per descrivere esperienze multisensoriali ed evocative, in cui sensazioni diverse si intrecciano richiamando anche ricordi ed emozioni. Nella degustazione del vino, questa dimensione multisensoriale gioca un ruolo fondamentale: le caratteristiche del vino possono attivare connessioni profonde con la memoria e l'esperienza individuale, evocando immagini, suoni o emozioni legate a momenti vissuti.

Due testimonianze letterarie significative possono aiutarci a comprendere meglio questa affermazione. Gaston Bachelard, nel suo *Il diritto di sognare* (1974), scrive:

> «Questo bicchiere di vino chiaro, fresco, secco, riassume tutta la mia vita campagnola. Le persone credono che io beva, ma io ricordo».

Marcel Proust, ne *Alla ricerca del tempo perduto* (1913-1927), intreccia ricordi personali, fatti storici, sensazioni gustative ed emotive attraverso un approccio diacronico e didattico agli eventi.

È iconico l'episodio nel quale, passando davanti alla casa degli Swann, Proust avverte il profumo delle madeleine e questo risveglia in lui i ricordi d'infanzia e il legame con la propria madre che preparava quei dolci. La sensazione olfattiva diventa una chiave per aprire il baule della memoria. Nel degustare un vino e nel presentarlo a un consumatore attento, dovremmo cercare di suscitare in lui, attraverso il meccanismo della cosiddetta "sindrome di Proust", il ricordo di una sensazione piacevole che ha provato in passato, evocando quei descrittori sensoriali presenti nella molteplicità dei cibi, dei fiori o dei suoni che accompagnano tutte le fasi della nostra vita.

Mineralità e sinestesia: quale rapporto?

Il termine "mineralità", un descrittore sensoriale sempre più utilizzato nella stampa, specie quella anglosassone, associa questa caratteristica a un'alta qualità, all'originalità e all'autenticità, soprattutto per i vini bianchi prodotti nelle regioni settentrionali come il Riesling. Esprime una sorta di "senso del luogo" che evoca il rapporto tra la composizione chimica (minerale) del suolo e il vino. Per alcuni, rappresenta una caratteristica distintiva di vini francesi quali Sancerre, Loire o Chablis ed è legata a vitigni come chardonnay, sauvignon blanc o chenin blanc. I vini italiani che meglio incorporano questo carattere sono Timorasso, Lugana, Greco di Tufo, Carricante dell'Etna e alcuni Verdicchio invecchiati.

Descrivere gli stimoli sensoriali che la mineralità suscita non è facile: la sua origine multisensoriale e multimodale combina olfatto, gusto e reazioni trigeminali. Una mineralità salina si distingue da una mineralità amara. Viene descritta con note aromatiche di selce, pietra focaia, pietra asciutta, muschio, grafite, ostriche fresche, fumé e cherosene (o *petroleum*) nei vini invecchiati.

La ricerca è fortemente scettica di fronte alle ipotesi fantasiose di una relazione diretta tra i minerali presenti nel suolo e le caratteristiche del vino, e attribuisce invece queste note sensoriali a composti di zolfo (tioli) presenti in forma ridotta, legati al vitigno e alle condizioni di maturazione delle uve. In effetti, il benzene-meta-etiolo conferisce una dimensione empireumatica al vino, molto simile alla nota di pietra focaia.
Le sensazioni gustative sono il risultato di stimoli eterosensoriali localizzati in diverse parti della bocca che trasformano i messaggi chimici in messaggi elettrici (trasduzione). Tutte le cellule gustative hanno recettori di tre categorie: recettori dell'acidità, canali ionici e recettori di membrana. Questi interagiscono tra loro ed è quindi illusorio pensare che esistano aree specifiche nella bocca deputate alla percezione del sapore minerale. I sensi del gusto non sono né analitici né sintetici: sono solo discriminatori e possono fare la differenza, per esempio, tra l'acido tartarico che evoca durezza e l'acido succinico che dà una sensazione salata e amara.
Perché allora l'attributo "mineralità" ha un tale successo nella comunicazione? Forse perché richiama un rapporto quasi fisico con un luogo e soddisfa il desiderio di ricercare le origini, la nostalgia per i ricordi giovanili e per le vacanze in campagna. Questo accompagna la mutazione del gusto che rifiuta i vini concentrati e legnosi, prodotti in un contesto economico consumistico con strategie di marketing aggressive, per volgere invece lo sguardo verso prodotti più equilibrati, provenienti da terroir più rispettati che sono in grado di rivelare questa mineralità grazie a una viticoltura più eco-sostenibile.
La percezione della mineralità, allora, oltre a individuare un certo equilibrio tra sapore e note aromatiche, non è il riflesso imperfetto di un suolo, ma una rappresentazione mentale che designa, con un sostantivo, un forte potere evocativo. La mineralità di un vino

può diventare uno stimolo di stampo sinestetico perché porta alla mente la naturalità del luogo e il ricordo di sensazioni passate.

La magia del racconto: l'attualità del mito

Nel paesaggio vitivinicolo italiano è ancora possibile incontrare la realtà del mito. È qui che entra in gioco la dualità *ctonio/epictonio.* in greco, *Khtohōn* significa "terra" e si ritrova nella parola autoctono, che significa "nativo di quella terra". Nella mitologia greca, le divinità ctonie come Gaia, Ade o Demetra si distinguono dalle divinità epictonie (creatrici della terra) tra le quali troviamo Urano, Zeus e Atena.

Qual è il loro rapporto con il vino e il paesaggio? Un vino rosso, di struttura, prodotto in un paesaggio aspro come la Valtellina o assolato come la Calabria, è ctonio: un prodotto della terra, con radici profonde, conservato nella terra, nelle profondità delle cantine e nelle anfore di terracotta. Durante il rituale della libagione del simposio, gli antichi greci versavano le prime gocce di vino sulla terra come tributo alle divinità ctonie. Uno spumante come il Prosecco, prodotto sulle dolci colline dai connotati romantici di Valdobbiadene, ha caratteristiche epictonie: non invecchia in cantina, è esposto alla luce con le sue bottiglie di vetro chiaro, è il vino delle feste all'aria aperta, testimonia la bellezza dei paesaggi e si adatta sicuramente meglio alle pratiche socio-culturali moderne.

Esiste anche un'altra dualità, quella dionisiaca/apollinea riportata ai giorni nostri da Nietzsche. Da una parte abbiamo Dionisio, il dio del vino e dell'eccesso. Beve il vino in purezza, senza prima versarlo nell'apposito vaso dove era consuetudine per i greci diluirlo con acqua e, per questo, è chiamato *acratophoro.* Egli identifica il vino come carnoso, forte, alcolico, strutturato: ancora una volta, il protagonista è il paesaggio selvaggio, con connotazioni boschive e rocciose. Dall'altra parte abbiamo invece il vino della temperanza, dell'elevazione spirituale e del banchetto greco, che non era un incontro di ubriachezza, ma un momento di discussione, narrazione, canto. Chi deve essere invocato, in questo caso? Apollo, *Phoebus*, il dispensatore di luce. Rappresenta il piacere di bere come un momento per lo spirito, non solo per la carne o per il desiderio di provare ebbrezza.

Tradizione e ricostruzione della memoria

Non è il *genius loci*, lo spirito di un luogo, ad animare un territorio vitivinicolo, ma il *genius saeculi*, lo spirito del tempo, il quale ci permette di narrare e di pensare. Il paesaggio è dunque uno spazio costruito, dove lo spirito del luogo non si trova solo nel simbolismo contemplativo dei padri fondatori e nelle divinità che li proteggono, ma deve essere valutato anche con lo sguardo del positivismo, dove l'opera dell'*homo faber* si integra in modo ammirevole con le caratteristiche naturali di un territorio.

Il neuromarketing, basato sullo sviluppo delle scienze cognitive, utilizza lo *storytelling* per attivare un processo di stampo sinestetico attraverso il quale uno stimolo visivo (il paesaggio e i suoi punti di riferimento) o un'evocazione culturale vengono associati alla percezione sensoriale del vino e al desiderio di ripetere l'esperienza gustativa, fornendo poi stimoli al pensiero.

Quando investighiamo il significato del mito per coglierne l'essenza o la verità, ciò che conta prima di tutto è la memoria collettiva con la quale un gruppo sociale ricostruisce il proprio passato, adattandolo ai contesti culturali del presente dai quali può poi plasmare il futuro. L'esempio della *Tabula Peutingeriana*, che raffigura la topografia dell'antica penisola italiana, ci permette di collegare i "luoghi" identificati sulla mappa con i vini (e i vitigni) del passato.

Nel significato del mito, il paradigma degli eventi epici è narrato come *fabula*: la tradizione deve essere utilizzata indipendentemente dal fatto che questi vini possano essere considerati antichi o recenti, originali o ricostruiti. Ciò che conta è che il racconto "mitico" sia presentato come un reperto della tradizione, ma soprattutto che sia considerato significativo per la comunità di riferimento.

Nei rapporti tra luoghi e mito, la questione è soprattutto topografica. I luoghi sono ancora pieni di "segni" che, in un passato *fabulosus*, trovano nei nomi lo strumento per documentare la propria origine in un intreccio indissolubile. La *fabula* e il luogo diventano realmente "favola". Si stabilisce un dialogo tra il posto e il racconto dove è il vino a prendere "senso" dalle antiche *fabulae*, le quali trovano il proprio fondamento nei territori in cui le viti selvatiche furono addomesticate e dove le loro uve vennero vinificate per la prima volta. Il mito è, dunque, il vino.

Il *topos* delle origini troiane della civiltà latina, attraverso i protagonisti dei *nostoi* omerici e virgiliani, non si esaurisce con il mito della fondazione di Roma e Napoli, ma diventa il mito paradigmatico della cultura rinascimentale, che identifica negli eroi greci i fondatori di molte città italiane. Padova è città antenorea; Mantova si ricollega a Mantuo, figlio del tebano Tiresia; Sulmona fu fondata da Solimo, compagno di Enea; Bari è legata al mito di Dedalo, fratello di Minosse.

Secondo l'antropologia strutturale, la socializzazione di un luogo o di una città si realizza spesso attraverso un oggetto di riconoscimento, collegando mito e linguaggio: una montagna modellata in una torre o un ponte che attraversa l'acqua, con il significato di connessione e unione. Qual è lo strumento per leggere in quel luogo il senso offerto dal mondo naturale ed è alla base del processo di trasformazione che porta dalla natura alla cultura, dallo stato selvaggio a quello civilizzato? Il vino, in quanto risultato dell'addomesticamento della vite selvatica, diventa una sorta di pietra miliare che scandisce il tempo e rinnova la memoria. In Italia, le comunità vitivinicole formano una cultura originale nel mondo agricolo. Pur fortemente ancorate alla realtà, esse hanno la capacità di reinventare perpetuamente il proprio passato attraverso un prodotto emblematico e secolare: il vino. Rispetto alla standar-

dizzazione di gran parte dei settori della società contemporanea, queste comunità occupano uno spazio culturale a metà strada tra il rurale e l'urbano, fra tradizione e innovazione.
Nel rapporto tra le comunità agricole italiane e il territorio, inteso dallo storico francese Fernand Braudel come "realtà sociale prima di ogni cosa", il vigneto rappresenta l'elemento strutturante, una sorta di centro rituale carico di implicazioni economiche, politiche e simboliche. Per gli antropologi, i centri rituali terreni hanno una consistenza materiale (un manufatto in legno o pietra, un luogo carico di significati simbolici come un vigneto) e svolgono una vera e propria funzione di ancoraggio al territorio, rappresentando un centro di aggregazione sociale con contenuti trascendentali.
Il mosaico delle viticolture italiane non si discosta molto da questa analisi: ogni realtà viticola è una sorta di crinale tra popolazioni e lingue diverse, che porta con sé i segni più profondi, sebbene meno visibili, della difesa della "piccola patria" e della sua indipendenza, a volte in conflitto con le comunità vicine. Il paesaggio è quindi inteso come monumento, luogo dello spirito. Questa accezione semantica è chiara nella parola latina *monumentum*, che racchiude la radice indoeuropea *men*, da cui *mens* (intelligenza, ragione) e il verbo *memini* (ricordare, pensare, riflettere).
Osservando una carta geologica della nostra penisola, colpisce la distribuzione apparentemente caotica dei colori. Essa raffigura le famiglie delle matrici geologiche, le unità stratigrafiche e le tipologie di roccia, testimonianza della grande e varia ricchezza dei suoli che l'hanno generata. L'Italia rappresenta infatti circa il 50% della diversità pedologica mondiale. Non solo: è anche il Paese più ricco di biodiversità vitivinicola, con circa 2300 vitigni, di cui 644 iscritti nel Registro Nazionale delle Varietà e quindi ammessi per la coltivazione. Se incrociamo questi dati otteniamo un calei-

doscopio di interazioni tra territori e varietà che fanno dell'Italia un vero e proprio scrigno di esperienze vinicole e culturali senza pari al mondo.

Lo studio dei paesaggi vitivinicoli, per il carattere fortemente identitario e per le complesse implicazioni connesse alla qualità del vino, è sempre meno orientato verso aspetti deterministici ed economici e più legato a contenuti culturali e organizzazione sociale. Inoltre, poiché il paesaggio è la relazione tra un soggetto spaziale osservato e un soggetto umano osservante, si crea un'interferenza insostenibile dal punto di vista dell'approccio analitico cartesiano al quale si ispira il nostro intero metodo scientifico. La scienza del paesaggio deve quindi essere intesa come una scienza della complessità e deve tenere conto di diverse sfere di conoscenza: scienze umane e materiali, affettività e razionalità, fisica dello spazio e metafisica dell'estetica.

Nell'immaginario collettivo, aspetti estetici e caratteri sensoriali appaiono spesso inconciliabili. Cosa hanno, invece, in comune? Storicamente, arte e scienza sono sempre state intrecciate; si pensi, ad esempio, al Rinascimento. A un certo punto, però, e loro strade si separano. Perché oggi l'arte conta di meno di un tempo? Perché la scienza è diventata meno accessibile? È difficile far dialogare le cosiddette "due culture" in un mondo che presta maggiore attenzione alla cultura umanistica rispetto a quella scientifica, pur essendo tecnologico.

Le cose non sono miracoli, il fuoco non viene portato da Prometeo. La scienza avanza attraverso un duro lavoro teorico e pratico, ma la cultura è ancora dei miti. La bellezza, valore fondante, può diventare l'elemento di fusione tra due ambiti disciplinari così diversi. I metodi della storia quantitativa possono essere il nuovo punto di riferimento per superare una visione eccessivamente antropocentrica dell'ambiente geografico. In questo modo, i proble-

mi dell'industria non vengono separati da quelli dell'agricoltura e si sviluppa il concetto di *agrosistema*, il cui studio richiede competenze specifiche nei campi dell'ecologia, della geografia degli insediamenti e delle scienze naturali.
L'analisi del contesto è fondamentale per comprendere le relazioni tra le società antiche e l'ambiente che le ha ospitate, in cui le microstorie sono contenitori reattivi in cui organizzazioni agricole, strutture insediative e forme architettoniche interagiscono in un complesso puzzle di componenti idro-geomorfologiche, climatiche e biologiche (i vigneti). Il muro di contenimento di una terrazza o di un vigneto, rispettoso della morfologia di una pendice, non impone un'aggiunta in sovrapposizione agli elementi ad essa precedenti, ma costituisce l'espansione e la crescita della realtà esistente.

L'importanza della retorica

Gran parte delle attività di un professore universitario o di un comunicatore consiste nell'interagire con gli studenti e con il pubblico. Sono preparati per farlo? Affatto, come osserva giustamente Flavia Trupia nel suo *Viva la retorica, sempre! Il superpotere della parola* (2024). Insegnare e comunicare richiedono una preparazione specifica che spesso manca. Bisognerebbe essere retori prima che docenti. Chi sa parlare bene vende meglio i prodotti, progredisce più facilmente negli studi o nella carriera accademica e cattura efficacemente l'attenzione del pubblico durante gli eventi. La retorica è, in definitiva, uno strumento di persuasione. Purtroppo, mentre nelle università americane è materia di studio comune, in Italia viene insegnata raramente.

Nel linguaggio moderno, il termine "retorica" porta spesso con sé un'accezione negativa: evoca discorsi vuoti, poveri di impegno intellettuale e improntati a un'artificiosa ricerca di risonanza, anche utilizzando luoghi comuni e ampollosi giri di parole. È questa la cattiva retorica talvolta riconosciuta nei politici. Esiste tuttavia anche una buona retorica, quella praticata da chi è in grado di uscire da sé stesso, concentrarsi sul proprio pubblico e ottenere il risultato desiderato attraverso le parole. Il buon retore osserva gli ascoltatori, ne interpreta emozioni, desideri e timori e modula il proprio messaggio di conseguenza.

È forse arrivato il momento di riconsiderare la retorica del vino. I linguaggi e le narrazioni di oggi usano molto frequentemente le medesime parole, omologandosi al punto di appiattire la percezione anche di prodotti molto differenti tra loro per qualità e reputazione. In un mondo del consumo profondamente cambiato sotto il punto di vista di generazioni, gusti, accesso e diffusione, lavorare sul linguaggio per diversificare la comunicazione è imperativo.

La modestia che conquista

Chi parla in pubblico non è immune dall'ansia da prestazione. Uno degli espedienti per superarla è l'*excusatio propter infirmitatem*, ovvero una dichiarazione esplicita della propria inadeguatezza da parte del locutore. Condividere questo disagio con il pubblico non solo aiuta il relatore a superare l'imbarazzo, ma trasforma il suo senso di inferiorità in un'opportunità per apparire più accessibili e autentici. Ammettendo di essere a disagio nel proprio ruolo, l'oratore si pone in una posizione di non-superiorità e accorcia la distanza tra sé e il pubblico, creando così un legame con esso.
Questa *excusatio* è impiegata dagli oratori di ogni tempo per evitare l'autocelebrazione, la quale risponde alla speranza di guadagnare credibilità e autorevolezza ma che, il più delle volte, sortisce l'effetto opposto. Un approccio umile, che è una sorta di *captatio benevolentiae*, è un efficace strumento per catturare le simpatie dell'uditorio e ispirare in esso una disposizione positiva all'ascolto.

Il canone retorico

Comunicare efficacemente è un'arte che richiede metodo e consapevolezza. Per costruire un discorso chiaro e persuasivo è utile seguire un modello strutturato, capace di guidare il pubblico in un racconto ben costruito e memorabile. Lo schema classico della retorica prevede cinque fasi.
Inventio: l'invenzione o sviluppo di una tesi o di un'idea, da dichiarare all'inizio e articolare all'interno del discorso. Il relatore individua gli argomenti a supporto della tesi, generalmente tre o quattro. Ad esempio, parlando di vitigni autoctoni, si potrebbe iniziare definendo l'autoctonia, proseguire con il rapporto tra viti selvatiche addomesticate e vitigni coltivati, evidenziare il contributo delle migrazioni al patrimonio genetico locale e concludere con il problema dell'erosione della biodiversità genetica e le strategie per contrastarla.

Dispositio: la scaletta delle argomentazioni da presentare nel discorso. Si inizia con un'*excusatio*, si presenta la tesi, si illustrano gli argomenti a supporto e si conclude con uno slogan finale o una citazione. È fondamentale evitare di sovraccaricare il pubblico con troppi concetti o immagini: risulterebbe in una dispersione sia dell'attenzione che del messaggio.
Elocutio: la cura dello stile espressivo. Il relatore sceglie la veste linguistica e lo stile più adatti. Il ricorso all'uso di metafore aiuta a rendere il discorso più coinvolgente ed è funzionale alla costruzione della storia.
Memoria: le tecniche per memorizzare il discorso, affinché l'oratore possa esporlo con sicurezza. Sebbene l'avvento di PowerPoint e relative slide abbiano alleggerito molto l'impegno mnemonico richiesto all'oratore, avere fissato in mente uno schema preciso e solido da seguire rimane essenziale.
Actio: gesti, intonazione e comunicazione non verbale che accompagnano l'esposizione. Il linguaggio del corpo, il tono della voce, il contatto visivo e i movimenti delle mani conferiscono intenzione e autorevolezza al discorso. Da non sottovalutare sono inoltre le pause strategiche, che contribuiscono a mantenere vivo l'interesse del pubblico.
Tra le frecce all'arco della narrazione, la figura retorica della metafora è la regina indiscussa. Come suggerisce Aristotele, però, sono di basilare importanza per la persuasione anche altri aspetti come *ethos* (la credibilità dell'oratore), *pathos* (la capacità di suscitare emozioni) e *logos* (l'abilità di far leva sulla logica e la razionalità del pubblico).
Di grande efficacia nella narrazione sono poi i luoghi della mente, i *topoi* della retorica aristotelica. Indicano un motivo, un elemento tematico ricorrente in uno schema narrativo come, ad esempio, i viaggi di ritorno – i *nostoi* – o gli eroi. Sono largamente usati dagli oratori perché rappresentano elementi comunicativi noti e compren-

sibili per il pubblico. Nella *Retorica*, Aristotele elenca diverse tipologie di *topoi* e si sofferma sull'efficacia dei "luoghi comuni", intesi dal filosofo non come le trite formule stereotipate o pregiudiziali dell'accezione moderna, bensì come massime generali e nozioni che non ricadono sotto il dominio di una scienza specifica e che, per questo, possono estendersi a ogni argomento. Tali *topoi* si definiscono comuni perché condivisibili da tutti, e il loro ruolo è di fornire un'idea collettiva e accessibile all'intero pubblico. Tra i luoghi comuni, di particolare effetto per la loro forza psicologica sono quelli incentrati sull'opposizione dei contrari.

Nel mondo del vino, gli esempi abbondano: tradizione contro innovazione, vini invecchiati in barrique o in grandi botti, vitigni autoctoni *VS* varietà internazionali, lieviti indigeni rispetto a quelli selezionati, viticoltura convenzionale e viticoltura biologica. È il caso però di domandarsi se enfatizzare queste contrapposizioni – che attirano il consumatore, soprattutto quello meno esperto e influenzabile – sia davvero utile per fornire una comunicazione trasparente e rassicurante sul vino. Il consumatore trae davvero vantaggio da una comunicazione che mette in competizione i vari aspetti della produzione, oppure è preferibile trovare un compromesso, come spesso accade nella pratica? Il rischio è quello di creare divisioni e tensioni, generando tribù di persone che difendono con fervore le proprie posizioni a scapito del consumo del vino stesso.

Invece, serve una comunicazione seria, ricca di contenuti tecnici ma espressi in un linguaggio accessibile, che metta il consumatore nella condizione di poter effettuare una scelta ragionata sulla base di informazioni chiare, non solo guidato dall'emotività. Bastano poche parole guida: paesaggio, territorio, storia, cultura e ambiente.

Un "nuovo mondo" fra tradizione e innovazione

Un modello narrativo efficace per raccontare un vino, un territorio o un brand è quello che mette a confronto passato (storia) e presente (attualità). La ricostruzione storica attinge spesso a simboli familiari, come le tracce della Grecia classica nell'architettura, i richiami mitologici ai vini del mito o l'immagine della viticoltura ancestrale evocata dalle viti ad alberello. La parte che dialoga con il presente, invece, pone l'accento sulla sostenibilità ambientale, sul riconoscimento internazionale dei vini del territorio o su pratiche innovative come la coltivazione di vitigni PIWI. Questo approccio dà vita al concetto di un "nuovo mondo" in cui storia, mito e contemporaneità si fondono in un vino moderno, pensato per intercettare i gusti della Generazione Z.

Un esempio emblematico di questo tipo di *storytelling* è la recente presentazione del nuovo Franciacorta Bellavista *Alma Assemblage 1*. Qui, una metastoria racconta la sfida di produrre un vino di altissima qualità, proponendo l'immagine di un mondo nuovo come promessa di equilibrio tra i valori di rassicurazione e di pace che identificano il brand (o il territorio). Ciò avviene attraverso il ricorso a simboli che evocano sentimenti universali come innocenza, creatività, armonia, audacia, amore.

Vittorio Moretti affida alla figlia Francesca, supportata da collaboratori d'eccellenza come Richard Geoffroy in cantina e Marco Simonit in vigna, il compito di affrontare la sfida del cambiamento climatico. Per farlo, coniuga tecniche innovative (come la *space economy*) con metodi tradizionali come la potatura conservativa e la vinificazione di uve provenienti da diversi terroir della Franciacorta. Il risultato è un vino che valorizza la biodiversità del territorio attraverso strategie sostenibili, in un delicato equilibrio fra tradizione e creatività, per interpretare ed esprimere le nuove tendenze del gusto.

Alle proprie competenze tecniche – si è laureata in Enologia all'Università di Milano – e all'esperienza come *Chef de Cave* in Champagne, Francesca Moretti associa la tradizione spumantistica del territorio che costituisce lo *stil de vie* del brand. L'archetipo evocato è il cosiddetto *innocent*, che esprime il desiderio di purezza, buoni sentimenti e semplicità. Evoca l'infanzia e il sapore fiduciosamente idillico della famiglia. Richiama il rinnovamento, l'ingresso in una sorta di terra promessa e un senso mistico di integrità. Definisce, in sostanza, una efficace sintesi di uso consolidato e creatività.
Come si inquadrano questi elementi evocativi nello *storytelling* del vino?Il ruolo dell'eroe è interpretato dal consumatore. Il valore è rappresentato dal vino stesso, mentre il territorio è la condizione, il mezzo attraverso il quale l'eroe può raggiungere il proprio oggetto di valore.

Il ruolo della metafora nello *storytelling* del vino

La metafora sostituisce un termine con un equivalente figurato per facilitarne la comprensione. Il linguaggio tecnico della comunicazione specializzata può risultare aggregante per un pubblico ristretto di "addetti ai lavori" che lo comprendono, ma contemporaneamente escludere e allontanare chi non ha familiarità con il settore e vi si avvicina per la prima volta. I giovani consumatori, ad esempio, si sentono tagliati fuori da una comunicazione del vino fatta di terminologia e concetti complessi dei quali non afferrano bene il significato.
L'uso delle metafore può aiutare a rendere il messaggio più chiaro e accessibile. Il linguaggio deve essere "comune ma preciso", semplice ma non semplicistico. Nel *Simposio* di Platone, dialogando con il ricco discepolo ateniese Agatone, Socrate illustra la grande funzione pedagogica della metafora del bicchiere mezzo pieno. Nello *storytelling* del vino, alcune delle metafore più comuni per sviluppare e rendere comprensibile la narrazione sono quelle dell'eroe, del confine, del viaggio e dell'angoscia. Attraverso queste immagini, possiamo raccontare le molteplici viticolture italiane, l'origine dei vitigni coltivati e le preoccupazioni legate al cambiamento climatico.

La metafora dell'eroe

La figura dell'eroe fa parte di uno schema narrativo canonico ampiamente utilizzato nella comunicazione americana e, in particolar modo, nella pubblicità di prodotti sportivi o per bambini. Si basa sul percorso di un protagonista che affronta tappe e ostacoli per raggiungere un premio finale. Questa struttura narrativa, familiare fin dall'infanzia grazie a fiabe, fumetti e film, è alla base della cultura popolare, da *Harry Potter* a *Star Wars*, da *Indiana Jones* a *James Bond*; anche *Topolino*, nel suo genere, è descritto come un eroe. L'importanza dell'eroe è stata ben sintetizzata dall'attore Mel Gibson:

> «Ho sempre scelto storie di personaggi straordinari perché è questo che il cinema dovrebbe fare: raccontare storie di eroi da cui trarre ispirazione».

Anche le campagne elettorali seguono questa logica, trasformando i candidati in brand e protagonisti di un intreccio narrativo. Devono saper comunicare emozioni – giocando su sentimenti come paura e speranza – e valori che rispondano ai bisogni degli elettori.

Non è difficile intravedere questo stesso schema nella storia dell'*Odissea* e nel viaggio di Ulisse attraverso il Mediterraneo, che appaiono molto attuali confrontati con le avventure di James Bond. In entrambi i casi, ad esempio, si incontrano donne misteriose e luoghi esotici: *jet set*, Montecarlo, Cortina e Las Vegas per l'Agente 007; l'isola di Eolo, l'Etna, Ischia, l'Ade e i campi Flegrei – dove risiedevano le divinità e i miti – per Odisseo. Il vino, lo Champagne Dom Pérignon per Bond e il vino della Tracia usato da Ulisse per ubriacare Polifemo e sfuggirgli, è presente in tutti e due gli

scenari e porta riferimenti a un consumo elitario. Nell'*Odissea*, il vino assume il ruolo di aiutante dell'eroe, un archetipo che richiama la "pozione magica" della tradizione narrativa. Come l'intruglio preparato dal druido per Asterix e Obelix, il vino di Tracia diventa uno strumento per accrescere la forza dell'eroe e aiutarlo nella lotta contro l'antagonista.
Un concetto simile è stato sfruttato dal marchio di bevande sportive Gatorade, che ha coinvolto gli sportivi Usain Bolt e Serena Williams come protagonisti della propria campagna pubblicitaria. Il brand ha raccontato la storia della nascita di Gatorade e la creazione del suo *Sport Science Institute*, che studia scientificamente gli effetti della bevanda sugli atleti.
Un aspetto tanto fondamentale quanto difficile da realizzare nella costruzione di una narrazione efficace è l'approccio multidisciplinare. Un racconto diventa affascinante quando riesce a combinare elementi culturali, antropologici, storici, geografici, biologici ed enologici.
L'analisi della catena del valore del Müller-Thurgau offre un ottimo esempio per individuare i temi sui quali applicare il modello narrativo dell'eroe, contestualizzandolo in un territorio preciso. La scelta di questo vino non è casuale. Al contrario, è paradigmatica perché permette di combinare tutti gli elementi della narrazione che devono interagire tra loro. Il Müller-Thurgau è prodotto in un'area ben definita, la Val di Cembra in Trentino, e possiede una forte identità legata alle caratteristiche alpine del territorio. Il suo profilo aromatico, facilmente riconoscibile, è particolarmente indicato per un pubblico giovane, col quale riesce a creare un legame immediato. Inoltre, ha una storia avvincente. Questo vitigno è un incrocio realizzato da un ricercatore svizzero, Hermann Müller, nel cantone Turgovia (Thurgau) della Svizzera tedesca, nella seconda metà dell'Ottocento. Si tratta di un periodo drammatico

per la viticoltura europea, minacciata dalle malattie giunte dal Nuovo Mondo. Il Müller-Thurgau presenta tutte le caratteristiche per costruire una storia coinvolgente: il ruolo della genetica, il contributo della biologia molecolare che ha corretto l'errore iniziale nell'identificazione dei "genitori" del vitigno, il legame con un ambiente alpino puro e incontaminato, i suoli di origine vulcanica e la facilità di riconoscimento del vino. Questi elementi permettono di sviluppare con facilità concetti chiave come originalità, naturalità, identità e tipicità.

In questo contesto, l'eroe è il ricercatore che ha creato il vitigno, l'antagonista è l'opinione pubblica dell'epoca, contraria alla genetica. L'alleato è la necessità di trovare vitigni più resistenti in un'epoca segnata dalla crisi della fillossera, mentre il premio sta nella vittoria del progresso sulla tradizione e nel successo ottenuto tra i viticoltori tedeschi, che hanno reso il Müller-Thurgau il vitigno a bacca bianca più coltivato in Germania.

L'eroe archetipico: Ulisse

Ulisse è il prototipo dell'eroe, il brand distintivo della Magna Grecia che traccia la rotta per i navigatori euboici. La sua impresa è rappresentata dalle difficoltà che incontra lungo il cammino. Il suo avversario è la natura ostile del mare e delle divinità che lo governano, teatro del suo viaggio avventuroso. Il conflitto è incarnato dalle guerre tra i diversi popoli del Mediterraneo, che lottano per la sopravvivenza delle proprie colonie. Il trauma è la violenza – ad esempio durante l'incontro con Polifemo – a cui Ulisse risponde con la forza del proprio ingegno e con il tesoro che porta con sé. Il premio, l'evento di verità, è il vino di Tracia, simbolo di Dioniso che aiuta l'eroe a superare le ostilità. Gli oggetti magici sono i doni ricevuti lungo il cammino, come il vino del sacerdote Maroni, padre di Calipso. Quest'ultima, come anche Circe, fa parte degli aiutanti, donne che Odisseo incontra e che nutrono per lui sentimenti contrastanti. Il matrimonio finale è la vittoria di Ulisse sulla prepotenza e l'arroganza dei Proci: il trionfo del bene sul male, dei valori positivi come l'amicizia, la solidarietà tra compagni, l'affetto del figlio. Nei poemi omerici, il vino è spesso protagonista di episodi significativi, veri e propri *topoi* narrativi che lo vedono come elemento unificante tra i personaggi, una sorta di pozione magica.

Non tutti gli eroi incarnano valori positivi. L'impresa di Giasone e degli Argonauti, partiti dall'antica città di Iolco alla volta della Colchide – l'attuale Georgia – alla ricerca del Vello d'Oro, si sviluppa su due livelli narrativi: quello orizzontale, che segue le prove affrontate dagli Argonauti, e quello verticale, che si immerge nei miti eroici precedenti della saga omerica e dei *nostoi*. Giasone è un eroe accidentale, spesso travolto dagli eventi, privo di controllo sulla propria sorte e oscurato dalle qualità semidivine dei propri compagni. Il suo unico vero talento è una certa avvenenza, che lui non esita a sfoderare in modo spregiudicato per salvarsi la pelle. In netta contrapposizione alla figura di Ulisse, il mito di Giasone trova successo proprio nella sua *normalità*: alla fine, è il fascino di ciò che è umano a prevalere.

Giasone, Georgia, Mar Nero, *kvevri*, vino ancestrale e naturalità rappresentano un mondo diverso rispetto a Ulisse con il Mediterraneo, la Tracia, i *pithos*, le anfore, il vino dolce e aromatico. Questi *storytelling* seguono due percorsi distinti: uno racconta la colonizzazione greca delle coste orientali del Mar Nero, l'altro quella del Mediterraneo occidentale. Sebbene entrambi gli eroi condividano alcuni tratti comuni, come la diffusione del mito del vino e il legame tra vino e divinità, i simboli e gli archetipi che incarnano sono profondamente diversi e coinvolgono valori universali come il coraggio, l'onestà, la generosità, l'amore e l'innocenza.

Alla pari del mito di Giasone, il recente "ritorno all'anfora" per il vino può trasmettere un'idea di semplicità e autenticità, proprio come l'umanità dell'eroe.
I grandi personaggi della storia possono diventare strumenti efficaci per la costruzione di un racconto. Chi ha lasciato un segno indelebile nel proprio campo, grazie al carisma e alle proprie azioni, diventa una figura su cui costruire una narrazione potente. Il viaggio di Ulisse attraverso il Mediterraneo è una metafora straordinariamente efficace per raccontare il percorso dell'eroe: un viaggio alla ricerca di sé stesso, in cui la scoperta finale avviene grazie a un tesoro o un evento di verità, con l'aiuto di una serie di alleati. Lungo il cammino, l'eroe deve affrontare prove e ostacoli, mentre il premio finale è il raggiungimento del proprio sogno. Oggi, tra gli eroi moderni possiamo annoverare scienziati e ricercatori. Le loro storie affascinanti possono aiutarci a raccontare il progresso scientifico come l'unico rimedio ai problemi che, due secoli fa, hanno messo in ginocchio la viticoltura. Quelle stesse sfide si ripresentano oggi, seppur in forme diverse, con la crescente richiesta di un'agricoltura più sostenibile e la necessità di reagire al cambiamento climatico.

L'eroe moderno: Mario Incisa della Rocchetta

La leggenda di Mario Incisa della Rocchetta, legata alla narrazione del Sassicaia sin dagli anni '80, ha avuto origine ben lontano da Bolgheri. La sua storia si intreccia con quella dell'Europa a cavallo tra il XIX e il XX secolo, attraversando alcune delle grandi calamità e innovazioni scientifiche di quel periodo. Possiamo considerare Mario Incisa un eroe moderno: la sua vicenda è di grande attualità perché, nella sua esperienza di agronomo e ambientalista, si possono individuare alcuni antidoti efficaci per affrontare le sfide del nostro tempo, come il bisogno di sostenibilità ambientale e le conseguenze del cambiamento climatico.

Tutto ha inizio nell'Olgiata, la tenuta in cui Incisa si trasferisce negli anni '30 e dove rimane fino al 1943. È questa il suo vero progetto, finanziato dai successi nelle corse del suo purosangue Ribot, ricordato come il cavallo più forte del mondo per velocità e temperamento. La tenuta gli era stata ceduta dalla madre, Eleonora Chigi, appartenente a una nobile famiglia toscana, dopo essere stata riscattata dai fondi americani di Clarice della Gherardesca, moglie di Mario Incisa. Qui nasce un'azienda zootecnica modello, con case coloniche per i dipendenti, un acquedotto e un caseificio. Diviene inoltre la sede dell'allevamento dei cavalli da corsa della celebre razza Dormello-Olgiata, che diede i natali a campioni come il capostipite Nearco e il già citato Ribot.
Rientrato a Bolgheri nel 1943, Incisa avvia la produzione di frutta e ortaggi destinati alla Svizzera e di bulbi di gladiolo per i mercati europei e americani. Alla morte di Ribot, nel 1972, le entrate del marchese crollano: è il vino a risollevare le sue sorti. E pensare che il grande Ribot, a Bolgheri, non è mai arrivato.

Non solo Sassicaia

Dalla *brand image* che domina il mercato negli anni '80 alla *brand story* che si impone a partire dagli anni 2000, le sequenze narrative sostituiscono progressivamente le campagne pubblicitarie. Non è più sufficiente rendere un marchio famoso o farlo conoscere a un vasto pubblico di consumatori anonimi: oggi è necessario creare un legame unico ed emozionale tra il brand e i suoi sostenitori.
Non bastano più il celebre viale dei cipressi, la bellezza del paesaggio o la qualità dei vini per raccontare la storia di Sassicaia. Occorre dare alla storia di quel vino e di quel territorio una nuova dimensione capace di toccare le corde emotive del consumatore, in accordo con i sentimenti che caratterizzano la contemporaneità come la sostenibilità ambientale, il rispetto della natura e dei

relativi ritmi, il valore della biodiversità; valori che Mario Incisa della Rocchetta ha realizzato prima all'Olgiata e poi a San Guido e che ha saputo descrivere con maestria nel suo libro *La terra è viva*, pubblicato postumo nel 1984. In questo libro, Incisa racconta quei principi che oggi ritroviamo nell'anima dei consumatori. La narrazione della sua vita di agronomo illuminato e curioso si rivela il miglior veicolo dell'ideologia del cambiamento, ovvero la consapevolezza di essere sempre sulla soglia di una trasformazione. Il saggio è un appassionato grido di denuncia contro gli effetti devastanti dell'aratura profonda sulla vita delle piante, pratica impiegata negli anni '30 per bonificare le paludi italiane con l'introduzione della meccanizzazione.

Con l'arrivo della meccanizzazione nella seconda metà del XIX secolo e l'adozione dell'aratro reversibile, capace di dissodare, rivoltare e appianare il terreno, il ruolo dell'agronomo attraversa una fase di profonda trasformazione di fronte ai progressi della fisiologia, della biochimica e della chimica agraria. Certamente, gli effetti drammatici delle tempeste di polvere (*dust bowl*) che, negli anni '30, colpiscono la parte occidentale degli Stati Uniti, uniti a quelli della Grande Depressione e delle pratiche di aratura intensiva, lasciano un segno profondo nella visione agricola di Mario Incisa della Rocchetta, contribuendo alla nascita di una sua nuova coscienza ambientalista.

Si tratta di una rivoluzione concettuale, che porta al passaggio da un'agricoltura fillocentrica, incentrata sulla parte aerea della pianta, a un'agricoltura rizocentrica, focalizzata sulle radici attraverso lavorazioni e fertilizzazioni mirate, ma in un'ottica conservativa, volta a evitare l'ossidazione della sostanza organica. Nel 1959, Incisa dona al WWF i 90 ettari del padule di Bolgheri, che diventa il primo rifugio faunistico d'Italia insieme a Burano. Nel 1966, lo stesso Mario Incisa della Rocchetta diventa co-fondatore e primo presidente di WWF Italia.

Il rinascimento del vino italiano nella storia dei SuperTuscan

Il merito del termine SuperTuscan va a Frank Pryal, per anni editorialista enologico del *New York Times*, che lo conia per definire gli ora famosi vini fuori dagli schemi. I primi a nascere, Sassicaia e Tignanello, sono frutto dell'intuizione di Mario Incisa della Rocchetta e Nicolò Antinori unita alla competenza enologica di Émile Peynaud e Giacomo Tachis. Prima di loro, questa nuova concezione di vino toscano diverso dal Chianti era avvolta nel mistero. Molte aziende avevano iniziato, quasi clandestinamente, a introdurre vitigni internazionali nel taglio del Chianti. È però solo nei primi anni '80, durante un convegno a San Felice a cui partecipò anche Luigi Veronelli, che alcune aziende con export in essere verso il mercato statunitense parlano pubblicamente di questi vini.

Il Chianti era in crisi. Dopo la fine della mezzadria e con l'introduzione della legislazione sulle DOC, la ricostruzione della viticoltura toscana – favorita dal *Fondo Europeo Agricolo di Orientamento e di Garanzia* (FEOGA) – prende una direzione poco felice. Materiale di propagazione del Sangiovese di qualità modesta, sesti d'impianto troppo larghi e rese eccessive rendevano impossibile garantire uve adatte alla produzione di un vino di qualità. I consumatori stranieri cominciano ad abbandonare il Chianti. Le grandi aziende iniziano a eliminare le uve bianche dal famoso uvaggio Ricasoli e a sperimentare l'apporto di vitigni internazionali come cabernet e merlot per dare al Sangiovese maggiore stabilità cromatica e un profilo aromatico più ricco, ricorrendo anche all'aiuto della barrique.

L'avventura di Mario Incisa della Rocchetta come produttore di vino nasce dal desiderio di brindare alle straordinarie vittorie del suo cavallo Ribot negli ippodromi di tutto il mondo con un vino

all'altezza dell'occasione. Era stanco di dover festeggiare con i vini dei suoi concorrenti, soprattutto bordolesi, così decide di piantare un vigneto sul versante orientale del castello di Castiglioncello, utilizzando vitigni toscani come Canaiolo, Sangiovese e Trebbiano. I primi risultati non sono un granché: i vini risultano spesso acescenti e Veronelli, suo mentore e consigliere, non esita a criticarlo duramente. Una fitta corrispondenza testimonia il confronto, a tratti aspro, tra i due.
Il caso, o forse l'incontro con l'allora consulente per Antinori Émile Peynaud (Mario era cognato di Niccolò Antinori, padre di Piero), lo convince a sperimentare vitigni diversi da quelli toscani. Tramite il conte Salviati di Migliarino, a Pisa, scopre i vini prodotti con uve cabernet in quella zona. Con alcune talee delle viti dei Salviati realizza il primo impianto in una zona chiamata Sassicaia, non lontana dal mare. Nel 1968, con l'aiuto di Tachis e l'incoraggiamento di Veronelli, si presenta a un prestigioso concorso a Londra: il suo vino di Sassicaia batte tutti i migliori *Château* bordolesi. Il successo si ripete poi nel 1985 in una competizione tra i più importanti Cabernet del mondo.
La novità sta nell'aver superato il vincolo della tradizione che imponeva l'utilizzo di vitigni toscani e, soprattutto, nella capacità di comunicarlo senza esitazioni o complessi di inferiorità, raccontando una storia affascinante nella quale si intrecciano audacia, imprese ippiche, tecniche innovative di coltivazione e vinificazione, sfide sui mercati internazionali più prestigiosi. E tanti aneddoti. Una vera storia di successo!

La missione viticola di Pierre Viala in America

Tra gli eroi moderni dobbiamo includere anche scienziati e ricercatori il cui lavoro, attraverso storie avvincenti, racconta i progressi della scienza viticola. Tra questi, Pierre Viala occupa una posizione di rilievo per il suo viaggio del 1887 verso gli Stati Uniti alla ricerca di viti selvatiche da usare come nuovi portainnesti per la ricostruzione della viticoltura europea prostrata dalla fillossera. Già dal 1874, le prime viti innestate su piede americano presenti in Francia mostrano serie difficoltà ad assorbire il ferro dal terreno. Il problema sta nell'incapacità di quelle specie selvatiche di tollerare l'elevata presenza di calcare nei suoli di due importanti regioni vitivinicole francesi, Champagne e Charente. È a questo che cerca soluzione l'allora ventottenne professore di viticoltura presso l'ANA di Montpellier.

Prima di iniziare le proprie ricerche sul campo, Pierre Viala si ferma una settimana a Washington presso il *Federal Geological Survey* per individuare le zone più simili a quelle dai suoli marnoso-calcarei della Francia. Scopre una corrispondenza in un'area compresa tra il Missouri e le Montagne Rocciose, dal Montana al Texas, la cui origine geologica risale all'inizio del Mesozoico, nel Giurassico. Come per la Champagne, il suolo si era formato dal riempimento di una depressione dovuta alla tettonica delle Montagne Rocciose, con l'accumulo di depositi marini e frane provenienti dai pendii montuosi.

Per sei mesi, dal 5 giugno al 3 dicembre 1887, Viala viaggia attraverso gli Stati Uniti concentrandosi infine sul Texas, dove incontra l'antropologo franco-messicano Jean-Louis Berlandier. È grazie a lui che riesce a individuare alcune zone del Texas centrale dove le viti "si mantengono verdi e vigorose nei suoli marnosi". Le condizioni del viaggio sono estremamente difficili: lunghe permanenze in luoghi ostili e isolati, temperature torride fino a

45° C, un paesaggio selvaggio abitato da animali pericolosi. Il suo resoconto di missione è ricco di osservazioni interessanti, ma soprattutto documenta il successo della spedizione: riesce infatti a portare in Europa numerosi vinaccioli di viti selvatiche del tipo *Vitis berlandieri*, fondamentali per la creazione di ibridi di portainnesti adatti agli ambienti meridionali e calcarei che guideranno lo sviluppo della viticoltura continentale.

Non è difficile riconoscere nella missione scientifica di Viala negli USA il modello narrativo dell'eroe. Non si tratta in questo caso di una metastoria, anche se elementi di avventura e rischio – come popolazioni indigene ostili, crotali, difficoltà climatiche e ambientali – sono ben presenti. Viala è l'eroe in cerca di sé stesso, molto spesso rappresentato in conflitto con l'antieroe che, in questo caso, è costituito dagli avversari dell'innesto. Gli ostacoli sono le dure condizioni del viaggio, mentre l'impresa è la scoperta di un tesoro scientifico – la *Vitis berlandieri* tollerante al calcare – resa possibile anche grazie all'aiuto di un compagno di viaggio, l'antropologo franco-messicano Berlandier. Il premio, infine, è il coronamento dell'impresa con la creazione di nuovi portainnesti.

La metafora del confine

La metafora del confine è fondamentale per raccontare come si sia sviluppata la civiltà del vino in Europa e, in particolare, le origini delle viti coltivate. I confini sono notoriamente luoghi di crisi, tensioni etniche e contrasti, ma anche di scambio, regolazione di sistemi differenti, innovazione. Si tratta di un paradosso culturale che può essere allo stesso tempo un luogo di separazione – con la creazione di quelle che vengono definite "culture parallele" – e di incontro e fusione, di convergenze.
I confini sono definiti da innumerevoli elementi. Oltre alle delimitazioni spaziali, rappresentate da ostacoli agli spostamenti come mari, catene montuose, deserti e fiumi, esistono quelle culturali, religiose e linguistiche. In uno scritto pubblicato nel 1954, riferendosi all'esperienza coloniale dei Greci in Italia, il filosofo Martin Heidegger scrive:

> «La delimitazione non è ciò in cui ogni cosa si arresta, ma ciò da cui una cosa inizia la sua presenza. Ciò che descrive è quindi quanto di omogeneo c'è al suo interno oppure ciò che differenzia l'interno dall'esterno?».

Per rappresentare qualcosa che separa e unisce allo stesso tempo, è necessario identificare una sorta di terra di nessuno, un luogo tra due spazi – non necessariamente geografici – ognuno dei quali è occupato da una società o una cultura distinte. Come regola generale, la cultura più forte ha un'azione invasiva e, nel suo movimento di espansione, preme sulle società vicine. Nel caso della vite, è l'Impero Romano che esporta il proprio modello vitivinicolo nelle regioni limitrofe.

Nell'uso comune non facciamo distinzione tra confine, frontiera e *limes*. Tuttavia, un confine è uno spazio condiviso (*cum finis*) dove due culture che vengono a contatto si integrano per dar vita a una nuova cultura; molte varietà di uva sono nate grazie a questi intrecci. All'opposto c'è la frontiera, dove due culture si incontrano, si fronteggiano ma non si fondono, restando in opposizione. Le espressioni di queste culture vitivinicole rimangono immutate nel tempo, permettendoci di tracciarne la relativa origine. Il *limes*, infine, è un limite fisico dove le differenze sono rappresentate da montagne, fiumi e mari. *Limes* particolarmente significativi sono le isole o le valli difficili da attraversare, che rappresentano "frontiere nascoste", luoghi di isolamento e conservazione. Gli antropologi chiamano queste espressioni culturali *edge effect*, effetto bordo.

L'Italia, con la sua posizione tra Occidente e Oriente, posta lungo il cammino di popolazioni che l'hanno attraversata nel corso dei secoli, con le innumerevoli manifestazioni culturali e produttive della sua viticoltura, è l'illustrazione più efficace dell'incontro tra diverse culture. Così erano i *limes* romani lungo il Reno e il Danubio o il caso di Termeno in Alto Adige (Tramin in tedesco), il cui nome deriva dalla parola latina *terminus*, che indicava un confine geografico e politico per i Romani durante l'espansione verso il resto dell'Europa. Anche la traduzione tedesca di Collio, *den Ecken* (l'angolo) nasconde la sua identità di luogo situato al margine, su un confine.

L'Italia è ancora ricca di queste frontiere: pensiamo ad esempio ai "confini nascosti" che separano la viticoltura latina da quella greca nell'isola di Ischia o da quella dell'enclave etrusca di Capua, circondata dai territori degli eritresi. Lo stesso si può dire della netta linea di demarcazione tra la viticoltura d'ispirazione longobarda a ovest di Bologna e quella bizantina fino al Mare Adriatico. Curiosamente, il confine tra la viticoltura della Slovenia e quella del Collio non presenta oggi caratteristiche significative di distinzione, a parte la lingua: la condivisione della ribolla gialla come uva identitaria dei due territori cancella la percezione delle eventuali differenze.

Nella ricerca degli elementi che permettono di identificare i cosiddetti "confini nascosti", si utilizzano vari indicatori. Tra questi ci sono la terminologia (ad esempio il termine *karax*, il palo della vite di Massalia, che distingue la viticoltura dell'approccio greco ad alberello da quella dell'antica Liguria, dove le piante fungevano da supporto per le viti), l'uso di diversi tipi di attrezzi per la potatura (le falci o il coltello da potatura) e per la vinificazione (come il castello e la nave per il trasporto di uva e vino in ambito padano), le tipologie di torchio.

Di grande interesse culturale è anche il confine rappresentato, nella regione del Salento, dalle differenti pratiche di inumazione: pochi chilometri separano un popolo che onora i propri morti con la cremazione da uno che li seppellisce in piedi, con il volto rivolto a est, in fosse scavate nella pietra calcarea. I primi spengono simbolicamente la pira con il vino, rompendo il *kylix* dell'offerta; i secondi non celebrano alcun rito. Anche i vitigni ancora presenti nei territori occupati da questi due popoli sono diversi, così come le tecniche di vinificazione.

La dicotomia del paesaggio, segnata da forme di allevamento ad alberello o ad alberata, ha permesso di evidenziare le peculiarità delle tipologie viticole (vitigni, caratteri dei vini) dei popoli che sono entrati in contatto. Ischia è esemplare in questo senso. L'isola è caratterizzata da due matrici geologiche: quella tettonica, frutto dell'emersione della terra dal mare, e quella vulcanica, espressione dell'intenso vulcanismo campano. Sulle zone di matrice tettonica si insediano i Greci, che utilizzano le argille dei *flysch* per realizzare anfore, mentre su quelle vulcaniche si stabiliscono Etruschi e Romani per lavorare il ferro che proveniva dall'Elba. Da quest'area con connotazioni linguistiche neolatine, alla fine del ciclo storico dell'Età del Bronzo, partono tutte le navi che trasportano ferro in Grecia per la fabbricazione di armi. Dove operano gli Etruschi, la vite viene allevata su falanghe, sostegni di castagno molto alti ispirati alla viticoltura paleo-ligure. Nella parte dell'isola abitata dai Greci, dove permangono anche substrati lessicali dell'Egeo e del Caucaso preesistenti, la vite è allevata ad alberello o a spalliera bassa, detta localmente "latina". A separare le due diverse forme di allevamento c'è una strada, ancora oggi esistente e che ha permesso di conservare miracolosamente intatti i substrati originali. Dal punto di vista metodologico, un vitigno antico – anche non più coltivato e presente come reliquia in una collezione – può costituire un efficace indicatore dell'esistenza di un confine culturale nella viticoltura europea.

Confini e origine dei vitigni coltivati

Nei paesi con un'alta immigrazione, come gli Stati Uniti, il fenomeno della ricerca delle proprie radici familiari è in forte crescita. Solitamente, questo interesse si manifesta nelle persone intorno ai sessant'anni quando, intravedendo il tramonto dell'esistenza e in tempi di fede incerta, si avverte il bisogno di riconnettersi con il proprio lignaggio, riaffermando che siamo più di un singolo individuo. In passato, la ricerca della storia familiare richiedeva lunghi e faticosi viaggi. Oggi, la mole di informazioni disponibili negli archivi comunali e parrocchiali, così come nelle banche dati del DNA dei parenti, è aumentata considerevolmente. Tuttavia, una visita ai luoghi degli antenati rimane essenziale per cogliere dettagli importanti e scoprire i territori dove abbiamo radici invisibili, che il tempo non può cancellare.

Per la vite, i primi anni del nuovo millennio fungono da spartiacque tra la conoscenza ampelografica dell'Ottocento, priva di strumenti oggettivi per il riconoscimento dei vitigni, e i risultati delle analisi sul DNA, che inaugurano una fase inaspettata di scoperte e squarciano il velo di mistero che avvolgeva l'origine dei vitigni. L'irruzione della genetica molecolare nel riconoscimento delle varietà ha portato alla revisione di molti miti, speculazioni favolistiche e idealizzazioni riguardanti la loro origine. Soprattutto, ha permesso di tracciare l'area geografica di provenienza di numerose varietà, ricostruire i percorsi che le hanno condotte fino a noi, identificare parenti anche lontani che hanno contribuito alla loro formazione e mettere in luce il ruolo decisivo della cultura e degli aspetti materiali della vita delle popolazioni nella selezione, protezione e diffusione della vite.
Analizzando il comportamento di queste varietà rispetto alle condizioni ambientali e ai vini che da esse vengono prodotti, è possibile comprendere la fase climatica in cui sono state selezionate, l'entità degli interventi enologici e le aspettative dei consumatori dell'epoca. Le sorprese non mancano. Accanto a genitori insospettabili e padri difficili da identificare, emerge con sempre maggiore chiarezza una rete di parentele estremamente articolata, nella quale sono molto pochi i vitigni capostipite attorno ai quali si è formato il patrimonio ampelografico europeo.
Le tracce da seguire per ricostruire questa storia, apparentemente contraddittoria, si basano su segni da decifrare con il supporto di due discipline diverse negli strumenti d'indagine, ma complementari nei risultati: l'antropologia e la genetica molecolare.
Per la viticoltura europea, i confini tra popoli e culture, spesso segnati da forti contrasti e conflitti, sono luoghi privilegiati per la nascita di nuove espressioni nella coltivazione della vite: da periferie, si trasformano in centri gravitazionali. I Greci, con la pro-

pria esperienza coloniale e commerciale (*emporion*), sono i primi a diffondere il mito del vino in Occidente. I Romani, tramite le conquiste militari, portano la viticoltura nelle regioni di confine dell'Impero, dove la domanda di vino da parte delle legioni è elevata e favorisce così la nascita di nuovi modelli viticoli.

Per una serie di circostanze climatiche e commerciali, la maggior parte delle varietà oggi coltivate in Europa ha avuto origine lungo i confini orientali. È stato un processo lento e reiterativo: il DNA di alcune viti provenienti dalle regioni orientali, dove le pratiche viticole come la potatura e la selezione genetica erano da tempo consolidate, si è mescolato con quello delle viti addomesticate a partire dal Neolitico lungo le grandi valli fluviali.

Un aspetto cruciale nella creazione di nuove varietà è il modo in cui i Romani coinvolgono le popolazioni locali. Invece di ridurle in schiavitù, le integrano come alleati concedendo loro, su autorizzazione del Senato, il diritto di coltivare la vite nei loro territori, prerogativa inizialmente riservata ai soli cittadini romani.

Significativo è il ruolo svolto ad esempio dal traminer, un vitigno selvatico addomesticato lungo il Reno che, attraverso l'incrocio spontaneo con una varietà orientale chiamata heunisch, contribuisce alla formazione di numerosi vitigni che ancora oggi sono fondamentali per la produzione di vini di qualità come pinot, sauvignon e riesling. Attraverso ulteriori cicli di incrocio successivi, queste varietà hanno a propria volta dato origine a cabernet sauvignon, merlot, carménère e altri.

Il contributo dell'antropologia nel racconto dell'origine dei vitigni

Lo studio sull'origine e la diffusione dei vitigni coltivati rappresenta un paradigma interpretativo fondamentale per comprendere le scoperte più recenti su un tema centrale nella storia della viticoltura: dimostrare come l'antropologia, con i suoi metodi, si riveli decisiva per risolvere questioni altrimenti indecifrabili, anche quando affrontate con gli strumenti più avanzati della ricerca viticola.

Se si parte dalla linguistica, il ragionamento appare più lineare. Nel mondo si parlano oggi oltre cinquemila lingue, tutte riconducibili a tre processi fondamentali. Il primo è la colonizzazione, attraverso cui gli antichi, penetrando in una nuova area, introducono la propria lingua. Il secondo è la divergenza, per cui la lingua parlata da un gruppo cambia nel tempo. Il terzo è la sostituzione, che si verifica quando un popolo adotta una nuova lingua da un altro gruppo umano. Questi tre processi permettono di delineare schemi suggestivi sull'origine delle lingue attualmente diffuse nel mondo.

La circolazione varietale della vite ha seguito questi stessi schemi di diffusione. Nonostante i diversi approcci adottati nello studio dell'identità della vite nelle diverse parti d'Europa, emerge un tratto comune: il continente, privo di confini rigidi, è da sempre un crocevia di migrazioni, interazioni, ibridazioni, contrasti e conflitti tra popoli, e ha tratto il proprio nutrimento essenziale proprio da questa complessità culturale e politica.

L'analisi delle relazioni di parentela tra i vitigni coltivati in Italia ha evidenziato un legame genetico profondo tra le popolazioni presenti, al punto da poter ricostruire il loro pedigree a partire da otto vitigni fondatori. La forte connessione genetica tra i vitigni di origine greca e quelli ancora presenti nelle regioni storicamente

appartenenti alla Magna Grecia non si deve solo alle colonie o ai commerci dell'epoca classica, ma anche al successivo apporto culturale bizantino e basiliano.
La situazione politica e sociale tra il V e il VI secolo d. C. porta a una grave crisi demografica nell'Italia meridionale, che viene affrontata – soprattutto lungo la costa ionica e il basso Adriatico – con intensi flussi migratori provenienti dal Medio Oriente, in particolare dalla Siria e dai Balcani. L'influenza bizantina si rafforza progressivamente, specie nelle strutture ecclesiastiche, determinando un processo di progressiva ellenizzazione nelle aree in cui si insediano eremiti e monaci orientali. A queste comunità va il merito di aver dissodato terre incolte, educato le popolazioni locali all'agricoltura e trasmesso il gusto artistico e la lingua dell'Oriente. In particolare, ai monaci di San Basilio si deve la riorganizzazione delle popolazioni in comunità agricole a forte carattere religioso, oltre all'introduzione di nuove specie animali e vegetali dai propri luoghi di origine come il baco da seta, il gelso e la canna da zucchero.
La Calabria è una delle regioni più coinvolte da questa migrazione bizantina. Ne sono testimoni, ancora oggi, i resti degli antichi vigneti di Bivongi, Stilo, Brancaleone e Rossano, arrampicati sulle ripide pendici delle colline per difendersi dalle incursioni arabe. La malaria e il latifondo dei litorali avevano già provocato lo spopolamento delle pianure: questa viticoltura non è più esclusivamente di origine magnogreca, ma anche bizantina. Le numerose varietà che ancora sopravvivono, seppur come reliquie, sono la testimonianza di quel periodo e dei luoghi dai quali provenivano i monaci basiliani. Alcune di queste varietà, come il mantonico bianco, furono certamente importate dalle attuali Bulgaria, Romania, Serbia e Croazia, risultando così diverse dalle viti dell'antica Grecia e delle coste del Mar Nero.

La diversità biologica della vite coltivata, frutto di millenni di selezione, è un'eredità preziosa lasciata dalla natura e dai nostri antenati, impossibile da ricreare in laboratorio. Una volta distrutto, questo capitale non puotrà essere recuperato e sarà perduto per sempre. Tale diversità non ha solo un valore biologico in quanto fase di un processo evolutivo naturale, sebbene guidato dall'uomo. Rappresenta anche una risorsa economica di crescente interesse per i consumatori, attratti dai vini ottenuti da vitigni "minori", e per le attività legate al turismo e alla gastronomia.
L'origine controversa dei vitigni può diventare un elemento chiave per interpretare l'identità di un territorio. La vite, infatti, può assumere un valore simbolico indipendentemente dal processo produttivo: è quindi essenziale capire perché una determinata varietà si sia sviluppata in una specifica regione e quale significato può esserle attribuito. Seguendo i principi dell'antropologia strutturale di Claude Lévi-Strauss, il mito diventa uno strumento metodologico per concettualizzare il problema dell'origine di un vitigno. Permette infatti di trasferire nel presente, in modo tangibile, la storia di una comunità.
Il mito dell'origine o domesticazione consente di spiegare la genesi, l'evoluzione e l'originalità produttiva per ogni tipo di vino. Le continue invasioni di popolazioni orientali in Europa sono vere e proprie rivoluzioni, caratterizzate da cambiamenti lenti ma costanti, proprio come la nascita e l'evoluzione di un vitigno. Anche il concetto di autoctonia perde progressivamente significato: i vitigni sono il risultato di un'intensa e antica circolazione varietale, che ha coinvolto anche aree molto distanti tra loro e spesso senza confini geografici definiti. Di conseguenza, il termine autoctono non identifica più un luogo preciso, ma piuttosto il momento in cui un vitigno si esprime al meglio attraverso le proprie caratteristiche produttive.

Grazie all'analisi sistematica del DNA nucleare mediante marcatori microsatelliti, è stato possibile individuare importanti collegamenti con il germoplasma viticolo di alcune regioni. Esempi emblematici si trovano in Veneto, Trentino-Alto Adige e Friuli-Venezia Giulia, dove la correlazione genetica rispettivamente di corvina e marzemino con il refosco dal peduncolo rosso ha rivelato antichi legami varietali. Allo stesso modo, la parentela genetica tra alcuni vitigni calabresi e siciliani con varietà greche o anatoliche conferma la continuità storica di queste interconnessioni.

La cultura come confine tra viti selvatiche e coltivate

Nel luglio del 1774, il botanico tedesco Samuel Gottlieb Gmelin muore in una zona selvaggia del Dagestan, dove è stato portato e imprigionato da un gruppo di montanari caucasici dopo averlo rapito. Nonostante la sua giovane età, ha già alle spalle lunghe spedizioni scientifiche nelle regioni intorno al Mar Caspio ed è responsabile della collocazione della vite selvatica nel contesto della classificazione sistematica linneana di *Vitis vinifera*. Lo studente che lo assiste nelle sue ultime ore riferisce che, nella sua agonia, lo studioso tedesco recitava i nomi latini delle nuove specie da lui identificate, tra cui – appunto – la vite selvatica.
A dire il vero, fin dai primi geoponici greci, la distinzione tra forme spontanee e coltivate di vite è sempre stata chiara. Il rapporto tra le due sottospecie era però ambiguo. Spesso, l'unico elemento che permetteva un'attribuzione tassonomica corretta era rappresentato dalla diversa epoca di domesticazione, che ha profondamente influenzato il fenotipo delle viti strappate da più lungo tempo alla condizione di piante selvatiche. Questa constatazione porta a rifiutare la categorizzazione della vite coltivata e di quella selvatica in due sottospecie distinte e a considerarle piuttosto due stadi cronologici dello stesso percorso evolutivo.

Gli sviluppi della biologia molecolare e le ipotesi sull'origine delle popolazioni umane hanno recentemente confermato che le varietà coltivate in Occidente non sono solo il risultato di fenomeni migratori, ma anche di fenomeni puntiformi di para-domesticazione che, in epoche molto remote, generazione dopo generazione, hanno agito sulle popolazioni selvatiche modellandone la struttura genetica. La metamorfosi della vite da pianta selvatica a pianta coltivata è il risultato esclusivo dell'attività demiurgica dell'uomo: nel lento progresso della selezione genetica e nel miglioramento delle sue caratteristiche qualitative, l'uomo si è rispecchiato, ha addomesticato sé stesso e ha instaurato un rapporto paterno con i vitigni.

A fronte di una variabilità genetica naturale – non sempre molto elevata – nelle popolazioni spontanee, la presenza di migliaia di vitigni nella viticoltura europea testimonia, accanto al ruolo importante dell'introgressione genica, l'essenziale contributo dei popoli che li hanno addomesticati, guidati dalla propria cultura e dai propri miti, a partire dalle scelte di vita quotidiana. Parafrasando un'affermazione dell'ampelografo Louis Levadoux (1956), il segreto delle origini della nostra viticoltura è nelle viti selvatiche. Nell'analisi degli individui ancora presenti nei luoghi della primigenia domesticazione e ancora frequenti nell'Italia centrale, confrontandoli con gli antichi vitigni di quelle aree, è forse possibile ricostruirne l'origine geografica, le fasi della coltivazione e la successiva migrazione in altri luoghi.

L'inizio – almeno quello letterario – della storia viticola del Mediterraneo occidentale in generale è contenuto nel Canto IX della già citata *Odissea*. Dalla descrizione dell'isola dei Ciclopi e dell'ubriachezza di Polifemo, appare chiaro che già a quel tempo esistevano due viticolture nel Mediterraneo. La prima, che ne interessa la parte orientale, produce vini "forti e scuri" come il vino di

Ismaro donato a Ulisse da Marone, sacerdote di Apollo. È proprio questo vino a far ubriacare Polifemo, che non era abituato a vini così forti. La seconda viticoltura è invece presente sull'isola dei Ciclopi,

> «(...) che nulla piantano con le loro mani né arano;
> tutto per loro cresce senza semina e senza aratura:
> grano, orzo e viti che producono vino da grossi grappoli, e la pioggia di Zeus le fa crescere».

In questo passo (Odissea, IX, 108-11) si descrivono chiaramente le caratteristiche di una vite selvatica, sottoposta a una tecnica di coltivazione primitiva o nulla.

La descrizione omerica non è tuttavia soltanto un documento – il primo in ordine cronologico – importante per comprendere come il mito del vino si sia diffuso nel Mediterraneo occidentale e quale ruolo abbiano avuto le saghe degli eroi fondatori e i viaggi di ritorno (*nòstoi*). Assume anche il valore di una metafora: quella della coesistenza di due modelli di viticoltura che sono sopravvissuti fino ai giorni nostri in molte aree d'Italia. Essi sono espressione di culture molto diverse, che hanno utilizzato vitigni e forme di allevamento distinti per ottenere vini ancora oggi facilmente distinguibili.

Sulla scia della nave di Ulisse, altri navigatori micenei e fenici nonché i primi coloni greci diffondono le tecniche di potatura e allevamento della vite, conosciute solo nel Caucaso e nel Levante, con vitigni di origine orientale. Si può immaginare che il testo omerico si riferisca alla produzione di un vino nei luoghi menzionati nell'*epos* dell'*Odissea*, sulla costa tirrenica (Golfo di Napoli), sulle Isole Eolie e nella Sicilia nord-orientale. In effetti, una parte della Lucania e della Puglia era già chiamata "terra del vino" o meglio "terra dei pali di vite" (*enotria*), a dimostrazione del fatto che la produzione di vino in queste zone è antecedente alla fondazione delle città protagoniste dei *nòstoi*. La vite era infatti presente nell'Italia meridionale prima dell'arrivo dei Greci, in forme di coltivazione cosiddette antropofile, soprattutto nei luoghi della civiltà villanoviana e dell'espansione etrusca. Con tutta probabilità, era il risultato o della domesticazione di viti selvatiche spontanee nei boschi planiziali, oppure di sporadici contatti con le popolazioni delle coste orientali dell'Adriatico e dello Ionio.

L'introduzione di nuovi vitigni nell'Italia romana è un fenomeno molto frequente, favorito dagli intensi contatti politici ed economici che essa intratteneva con i popoli orientali. L'obiettivo è di imitarne i vini, molto apprezzati soprattutto in epoca imperiale. Le nuove varietà si diffondono con disinvoltura anche grazie ai principi che ispiravano l'opera dei georgici latini, per i quali la qualità del vino non era espressione delle caratteristiche genetiche del vitigno, bensì delle condizioni ambientali.

Tra gli eventi che favoriscono l'arrivo di vitigni da altri luoghi è bene menzionare anche la sollecitazione dell'imperatore Probo, la cui moglie di origine aquitana Faustina chiede, a nome dei propri concittadini, l'annullamento dell'editto emanato circa duecento anni prima dall'imperatore Domiziano che limitava la coltivazione della vite al di fuori dei confini d'Italia. Il caso vuole che le legioni più fedeli a Probo, incaricate di ricostruire la viticoltura lungo i *limes*, fossero stanziate in Pannonia, il ventre molle delle difese contro le invasioni, dove esistevano vitigni altamente produttivi e resistenti al freddo. Molte varietà senza nome vengono trasportate lungo le rive del Reno e del Danubio per essere coltivate insieme alle viti spontanee, che erano già da tempo sottoposte a processi di proto-domesticazione e che avevano introdotto le popolazioni locali al vino, sebbene di qualità modesta. Queste

varietà vengono genericamente chiamate *heunisch* (o *hunnisch*, da Unni) o *gouais* (nella grafia francese con un significato dispregiativo) poiché provengono da territori abitati dagli Unni. Lo chardonnay ha la sua origine remota in questo contesto.
Non c'è stata progettualità nell'intervento di Probo. Agisce senza rendersi conto di ciò che stava facendo fare alle proprie legioni. Se avesse scelto di portare viti da altre regioni dell'Impero o se sua moglie fosse stata originaria della Dacia anziché dell'Aquitania, la storia della viticoltura europea avrebbe preso un'altra direzione e i vini che oggi rallegrano le nostre tavole sarebbero certamente molto diversi.

Lo *storytelling* dell'autoctonia: perché è necessaria una nuova narrazione

Il termine "glocalizzazione", coniato da Roland Robertson nel 1992, definisce l'articolazione ambivalente tra il globale e il locale, tra l'interdipendenza dell'economia mondiale da un lato e la dimensione particolare dei contesti locali – con le relative tradizioni – dall'altro. Queste due realtà sono allo stesso tempo contrastanti e complementari e riguardano non solo un insieme di attività economico-finanziarie, ma anche un complesso di codici simbolici, pratiche sociali e abitudini culturali legate alle comunità territoriali.
La produzione agroalimentare, e quella del vino in particolare, rappresenta il paradigma interpretativo per eccellenza di un bisogno di identità e riconoscimento, nato dalla paura di legami sociali anonimi in un mondo sempre più globalizzato. In questo contesto, la tradizione – che sia reinventata o mitizzata – diventa un ancoraggio emotivo e di appartenenza, come la definisce lo storico francese Jacques Le Goff. Il ritorno ai vitigni autoctoni e ai prodotti tipici della cucina italiana è una prova di questa tenden-

za, ma è necessario mettere in luce e allontanare alcune forzature. L'Italia è il secondo produttore mondiale di pomodori, dopo gli Stati Uniti. Il pomodoro e i suoi derivati sono ingredienti fondamentali della cucina italiana e un simbolo del Made in Italy, tanto radicati nella nostra tradizione culinaria da essere percepiti come un prodotto autctono del nostro Paese. Eppure, il pomodoro è arrivato in Italia solo dopo la scoperta dell'America. Una considerazione simile vale per i vitigni coltivati nel nostro territorio: la maggior parte di essi proviene da altri paesi e, spesso, ha dato origine a nuove varietà attraverso incroci spontanei. Solo una piccola parte dei nostri vitigni è il risultato della domesticazione delle viti selvatiche locali.

La circolazione varietale in Italia è un fenomeno molto antico e legato alla posizione baricentrica del Paese nel Mediterraneo, tra Oriente e Occidente. Le prime tracce della presenza di vitigni "stranieri" risalgono alla cosiddetta quarta migrazione delle popolazioni caucasiche, avvenuta durante l'Età del Bronzo Antico (tra il 3.000 e il 4.000 a.C.) probabilmente in seguito al disastroso evento noto come il diluvio del Mar Nero. Tuttavia, il contributo varietale più significativo si è verificato nel primo millennio a.C., con la colonizzazione greca dell'Italia meridionale.

L'uso dell'aggettivo "autoctono" per descrivere le varietà di vite è relativamente recente ed è il risultato di un atteggiamento culturale chiamato indigenismo, un orientamento dell'antropologia e dell'archeologia che tende a considerare ogni società come autonoma, analizzandone gli sviluppi in termini essenzialmente endogeni. Oggi definiamo "italiani" molti prodotti alimentari che hanno acquisito questa identità solo in tempi relativamente recenti o che hanno subito profonde trasformazioni strutturali e sensoriali a causa delle moderne tecniche di lavorazione; per non parlare di frutta e ortaggi che, spesso, di italiano hanno forse solo

la dicitura in etichetta. Tornando al vino, e in particolare a quello prodotto nel sud Italia, è evidente che l'uso di tecniche moderne per renderlo profumato di fiori e frutta, per conferirgli un'acidità che naturalmente non possiede o, peggio, per dargli una struttura più ricca e concentrata lo priva di quell'identità che lo rende riconoscibile e inimitabile.
L'importante variabilità genetica della viticoltura italiana, sebbene economicamente marginale, affonda le radici nel tradizionalismo contadino, una scelta obbligata per le comunità rurali del passato che dovevano affrontare, con strumenti inadeguati, il tipico rischio derivante dall'imprevedibilità delle condizioni ambientali. Dove si vive ai limiti della sussistenza non c'è spazio per la sperimentazione, tanto meno per quella genetica, poiché comporterebbe rischi dalle conseguenze potenzialmente catastrofiche. Detto altrimenti, l'evoluzione della piattaforma ampelografica è stata solo di recente il risultato dell'intreccio tra comportamenti economici e culturali. Le principali modifiche genetiche sono invece avvenute soprattutto per cause indipendenti dalla volontà dei viticoltori e sono spesso ricollegabili a grandi cambiamenti climatici.
Alcuni aggettivi attribuiti ai vitigni – ad esempio minore, locale, antico, autoctono, vecchio, domestico o indigeno – vengono spesso utilizzati impropriamente come fossero sinonimi. In realtà, un vitigno locale non è necessariamente autoctono, così come un vitigno minore non è per forza antico. Il ritorno sulla scena globale dei vitigni antichi o autoctoni ha richiamato l'attenzione sul ruolo cruciale della varietà nel racconto del vino. Siamo però in un'epoca in cui non c'è più tempo per approfondimenti geografici o per analizzare il fascino legato all'origine e alla storia, il tutto a vantaggio di un consumo guidato più dal cervello che dal palato. Di questi vitigni spesso ricordiamo solo il nome, magari perché

ci suona curioso o vernacolare: dimentichiamo le sensazioni gustative dei rispettivi vini, spesso prodotti con tecniche enologiche non coerenti con le caratteristiche del vitigno e del luogo forse sperduto in cui vengono coltivati.
Cosa dovremmo fare, allora? Quale direzione prendere in un periodo di grandi cambiamenti nei consumi, negli stili di vita e – non ultimo – nel clima? L'elemento pervasivo in ogni scelta, sia essa produttiva o di consumo, è la globalizzazione. Accolta da molti per il suo contributo alla complessità culturale e ai rapporti commerciali, da altri è vista come un fenomeno inquietante e pericoloso, da respingere. Tra produttori e consumatori è opinione diffusa che la globalizzazione sia la principale responsabile dell'appiattimento delle caratteristiche sensoriali dei vini prodotti in tutto il mondo, condizionati da modelli di perfezione che sfociano in banalizzazione. In questo senso, è fondamentale il ruolo della comunicazione e delle guide vini, incapaci di valorizzare la "bellezza dell'imperfezione": lungi dall'essere un difetto tout-court, quest'ultima è spesso la base di un vino fuori dagli schemi, un vino innovativo.
La globalizzazione ha però agito anche in senso opposto, creando nuove opportunità per forme alternative di viticoltura, come la biodinamica, nel tentativo di preservare manifestazioni della produzione vinicola che le pressioni del progresso scientifico tendono a stravolgere. Tuttavia, affidarsi a filosofie *new age* per produrre vino non ha molto senso, così come rifugiarsi in una tradizione vista come antidoto ai mali della ricerca. Nulla è più lontano dalla realtà: come affermano gli storici E. J. Hobsbawm e T. Ranger ne *L'invenzione della tradizione* (1987), le tradizioni sono sempre state create e reinventate per servire gli interessi di chi voleva legittimare il proprio potere.
Non è raro che una varietà locale sia altresì antica: anche lad-

dove sia problematico stabilire con certezza una scala temporale e individuare il momento esatto della sua introduzione in quel territorio, facilmente il vitigno viene coltivato lì da molto tempo. Una varietà molto antica, però, può non essere locale: il caso di vitigni provenienti dall'Oriente come moscato e malvasia, coltivati da secoli in territori lontani dal relativo luogo d'origine, ne è un esempio.

Spesso, le difficoltà nel ricostruire l'origine di un vitigno derivano anche dal cambiamento che il suo nome subisce attraversando paesi diversi e adattandosi ai dialetti locali. Molti vitigni assumono nomi che derivano da caratteristiche morfologiche o produttive come il colore dell'acino, il portamento della vegetazione, la forma del grappolo, il periodo di maturazione o il luogo d'origine. La selezione di questi vitigni, siano essi di origine locale o introdotti, è infatti fortemente integrata con il sistema agrario tradizionale e oggetto di pratiche colturali particolari.

La grande variabilità genetica intravarietale ha reso i vitigni capaci di adattarsi facilmente a diverse condizioni. Tuttavia, tale adattamento da solo non basta a mantenere il loro valore culturale: se un vitigno viene coltivato senza tenere conto del contesto in cui si è sviluppato, ossia senza considerare le tradizioni agricole e alimentari locali, le necessità del territorio e delle persone che lo coltivano e consumano, allora perde il suo significato come parte del patrimonio agricolo e culturale di quella società. In altre parole, un vitigno non è solo una pianta, ma anche il risultato di un equilibrio tra natura e cultura. Se lo si isola da questo equilibrio, si perde una parte della sua identità storica e tradizionale.

Se il concetto di autoctonia è più legato allo spazio (ovvero all'essere nativo di un determinato luogo) che al tempo (da quanto tempo quel vitigno viene coltivato in quel luogo), allora viene spontaneo chiedersi: quanto tempo deve passare prima che un vitigno possa

essere considerato espressione autentica di un territorio? È quindi lecito chiedersi cosa significhi realmente il termine "autoctono" in relazione a una varietà di vite.
Se ci attenessimo strettamente al suo significato etimologico di "vitigno nato in quel luogo", dovremmo dire che in Italia abbiamo non più di una dozzina di vitigni davvero autoctoni. Il termine si potrebbe infatti applicare solo ai vitigni derivati dalla domesticazione di viti selvatiche locali – come il lambrusco di Sorbara, il lambrusco a foglia frastagliata, l'enantio, il refosco, l'asprinio e il savagnin – o alle varietà nate da fenomeni di introgressione genetica come i vitigni dell'antica Pannonia o il *gouais*, alla base di molte varietà dell'Europa continentale. La maggior dei nostri vitigni proviene da altri paesi o è il frutto di incroci spontanei. Sembra dunque impossibile dare una definizione univoca di "autoctono" per la viticoltura dell'Europa occidentale: sarebbe forse più corretto sostituire il termine "autoctono" con "antico", "locale" o "italico".
È davvero necessario conoscere l'origine geografica di un vitigno per definirlo autoctono? Il Sangiovese, per esempio, discende dallo strinto porcino, un antico vitigno dell'area vesuviana, ed è imparentato con visparola e ciliegiolo, suo discendente. Ha legami di parentela con cultivar di quattro regioni italiane: Basilicata, con stampacavallo e strinto porcino; Calabria, con castiglione di bova, gaglioppo e toccarino; Puglia con negrodolce; Sicilia, con nerello, arbanello, frappato, lievuso, lucignola, orisi e perricone. In quale di queste regioni può essere considerato autoctono? Il punto è che il senso ultimo di questo termine non è quello di indicare il luogo d'origine, bensì il territorio in cui il vitigno esprime al meglio il proprio potenziale genetico. Ecco perché il Sangiovese è considerato nativo della Toscana, di Montalcino, del Chianti, della Maremma e di Montepulciano.

L'epoca dei vitigni apolidi, le cosiddette varietà internazionali coltivate ovunque e sempre più destinate alla produzione di vini per la fascia più bassa del mercato, è finita. Con essa si esaurisce anche la confusione generata da una pletora di denominazioni di origine senza tradizione né meriti qualitativi. Qual è, dunque, il futuro della viticoltura italiana?
Nella determinazione della qualità di un vino, la *querelle* ancora irrisolta sulla supremazia del vitigno sul territorio – e viceversa – è ormai parte integrante della storia della viticoltura. Fino all'Illuminismo è il luogo di produzione a prevalere sulla varietà. Nel 1868, alle soglie della crisi fillosserica, Jules Guyot rompe questo assioma con la rivoluzionaria osservazione

> «Il genio del vino sta nel vitigno».

A chi invoca il ritorno alla tradizione nella produzione vinicola si può rispondere che il modo più efficace per onorarla è quello di un "tradimento fedele". Il paradigma interpretativo che ha guidato il progresso vitivinicolo nei secoli è sempre stato l'innovazione genetica, rappresentata dalla circolazione varietale e dalla selezione di incroci spontanei o intenzionali. Dopo circa vent'anni di grande interesse per i vitigni autoctoni da parte di aziende viticole, ricerca e consumatori, i risultati economici ottenuti sono tutt'altro che entusiasmanti. Certo, esistono eccezioni: alcuni vitigni e relativi territori hanno raggiunto buoni risultati in termini di valorizzazione, ma molti altri rischiano l'oblio. È quindi necessario analizzare i fattori in gioco per meglio indirizzare gli sforzi futuri.
In sintesi, dei più di mille vitigni autoctoni italiani, quelli di reale qualità – ovvero quelli in grado di produrre un vino adatto ai gusti del consumatore moderno – sono pochi, forse non più di un centinaio. Ciò non toglie valore agli altri, che restano fondamen-

tali per la conservazione della biodiversità e per i programmi di miglioramento genetico. Solo poche varietà, sostenute da DOC e DOCG di successo come nebbiolo per Barolo, Barbaresco e Valtellina, o Sangiovese per Brunello, Chianti e Nobile di Montepulciano, stanno vivendo una vera espansione. Alcuni vitigni antichi sono stati valorizzati non grazie alle istituzioni, ma per la passione di pochi viticoltori: menzioniamo ad esempio Walter Massa per il timorasso, Elisabetta Foradori per il teroldego e Marco Caprai per il sagrantino. Pochissimi vitigni, infine, sono stati recuperati dall'oblio e inseriti nel Registro Nazionale delle varietà autorizzate alla coltivazione, condizione *sine qua non* per poter essere moltiplicati e piantati.

Vitigni internazionali e storia della viticoltura moderna in Italia

La presenza dei cosiddetti vitigni internazionali è talmente importante nella viticoltura mondiale che i consumatori non li considerano una categoria particolare: si sono radicati così profondamente nella cultura enologica da modellare, senza che ne siamo del tutto consapevoli, le immagini e le sensazioni che abbiamo in mente di un Chardonnay o di un Cabernet Sauvignon.

La definizione di vitigni internazionali nel linguaggio della viticoltura, non solo italiana, è piuttosto recente. Risale agli anni '80, quando in Italia fu avviato un grande progetto per recuperare e valorizzare le centinaia di varietà autoctone presenti, ma praticamente sconosciute, in tutto il Paese. È proprio dal contrasto quasi competitivo tra queste due grandi tipologie varietali che emergono le loro differenze compositive, il diverso contributo alle caratteristiche organolettiche dei vini e una presenza che in precedenza era stata quasi ignorata.

La circolazione varietale è un fenomeno molto antico nella viticoltura europea e, se si esclude il contributo della Grecia antica, si può dire che questa invasione silenziosa abbia avuto inizio nel Medioevo, con l'introduzione di alcune varietà adatte alla produzione di vini dolci (malvasie, moscati, vernacce) da parte di Veneziani e Genovesi, e sia poi proseguita con l'apparizione di alcune varietà catalane e spagnole (grenache-cannonau, forse vermentino) in alcune regioni del Centro-Sud. All'inizio del Settecento si registra il primo interesse per i vini francesi, in Toscana per la produzione del "claretto alla franzese" e in Friuli con l'introduzione del pinot nero. Sono tuttavia i Medici, nel 1773, a importare un gran numero di vitigni da diversi paesi europei per valutarli nelle proprie fattorie. Il giudizio sul valore reale di queste varietà, però, rimane critico.

Nel XIX secolo, la presenza di vitigni chiamati "stranieri", provenienti da altri paesi tra cui la Francia *in primis*, diventa importante dal punto di vista del loro ruolo nella qualità dei vini italiani. Nel 1903, Salvatore Mondini pubblica *I vitigni stranieri da vino coltivati in Italia*, una rassegna completa della presenza nelle varie regioni della penisola delle viti che, nel corso del XIX secolo, iniziano a svolgere un ruolo economico e non sono più solo presenti nelle collezioni ampelografiche. Il principale motivo di queste introduzioni, iniziate da alcuni importanti produttori vitivinicoli dell'Italia centro-settentrionale, è l'emulazione del successo mondiale dei vini francesi, all'epoca la più importante potenza economico-militare in Europa. L'obiettivo è alimentare un flusso di esportazione di vini imbottigliati per sostituire il commercio di vini assemblati, molto apprezzati nei mercati esteri ma poco competitivi dal punto di vista economico.

Nell'Ottocento, anche sulla scia delle scoperte scientifiche avvenute nel secolo precedente, la Francia diventa il paese leader nella produzione vinicola e nello sviluppo di nuove tecniche viticole.

Queste iniziano a permeare il mondo produttivo italiano dopo la seconda metà del XIX secolo, con l'Unità d'Italia, sia per il ruolo che la Francia di Napoleone III aveva svolto nelle Guerre d'Indipendenza, sia per la crescente necessità di rinnovare la viticoltura dopo le devastazioni causate dalla fillossera.
Le prime innovazioni trasferite dalla Francia, qualche lustro prima dell'arrivo delle cosiddette "malattie americane", riguardano alcuni vitigni – perlopiù di origine bordolese – importati da alcuni proprietari terrieri toscani e veneti per imitare quei vini "franzesi", i *Claret*, che riscuotevano tanto successo in Inghilterra.
Le prime collezioni ampelografiche, espressione del collezionismo ottocentesco e della nascita delle scuole di enologia, contribuirono alla conoscenza del comportamento di questi vitigni negli ambienti italiani e alla corretta vinificazione delle loro uve.
I primi impianti di cabernet sauvignon risalgono al 1820 e vengono realizzati nei pressi della città di Alessandria, in Piemonte. Le più importanti famiglie nobili del Piemonte, della Toscana e del Regno delle Due Sicilie, spesso imparentate con la corte francese e consumatrici esclusive di vini d'Oltralpe, iniziano a produrre vini da questi vitigni già dalla fine del XVIII secolo. In Italia non si parlava mai di vini destinati ai mercati esteri. In questo periodo, la neonata Accademia dei Georgofili stimola, nell'ambito di un concorso, la pubblicazione di opere di erudizione enologica per migliorare la qualità dei vini toscani e favorirne l'esportazione. Il vincitore, il medico Cosimo Villifranchi, denuncia con amarezza che, sia in Toscana che altrove, la qualità dei vini ispirati al modello francese non aveva alcun legame con i celebri vini dell'antichità italica, soprattutto a causa di una tecnica enologica inadeguata:

«Nel Chianti uve nobili, vini plebei».

Nelle istruzioni che fornisce per produrre il miglior Chianti possibile, però, lui stesso ricade nella tradizione delle lunghe macerazioni con le vinacce, responsabili dell'acescenza dei vini.

1825: Giuseppe Acerbi, scrittore e naturalista mantovano considerato il primo ampelografo italiano, compila un catalogo con ben 903 varietà di viti straniere presenti nella sua collezione, provenienti in gran parte dall'Impero Austro-Ungarico di cui la Lombardia faceva allora parte. Nel 1852, il marchese Incisa della Rocchetta crea un'importante collezione a Rocchetta Tanaro per la quale non si limita a raccogliere viti straniere, ma ne valuta anche il valore viticolo ed enologico. Cabernet, verdot, gamay e pinot si distinguono per qualità. Queste collezioni sono le prime responsabili della diffusione dei vitigni alloctoni in Italia, suscitando curiosità e voglia di sperimentare in molti viticoltori.
Cosa ha motivato i viticoltori a introdurre queste nuove varietà? Innanzitutto, il desiderio di migliorare la qualità dei propri vini, che riconoscevano essere inferiore a quella dei vini di altri paesi – soprattutto la Francia – e che non riuscivano a incrementare con le sole varietà autoctone. Inoltre, la necessità di estendere la coltivazione della vite a zone meno favorevoli, dove i vitigni locali non davano buoni risultati. Infine, la speranza di coltivare viti più resistenti alle malattie crittogamiche (come peronospora e oidio, comparse di recente) e alle avversità atmosferiche.
Il cambiamento accelera sensibilmente nella seconda metà dell'Ottocento con l'arrivo della fillossera, che rende necessaria la ricostruzione della viticoltura italiana su piede americano e offre l'occasione per diffondere quei vitigni stranieri che avevano già dimostrato buoni risultati negli anni precedenti. La già citata affermazione del medico e agronomo Guyot, che vede nel vitigno il genio del vino, contribuisce al processo, smentendo duemila anni di cultura georgica che vedeva nel terroir il fattore più determinante della qualità di un vino.
Il drammatico rapporto che si instaura tra la viticoltura europea e le malattie americane, alle porte del XX secolo, è alla base della

nascita della cosiddetta viticoltura moderna. Cambiano i luoghi della viticoltura: i vigneti specializzati, ordinati in filari rigogliosi, sostituiscono i campi arborati, dove si alternavano viti maritate a gelsi o aceri e terreni coltivati a prato o cereali. Cambiano soprattutto i vitigni. In particolare, vengono definitivamente abbandonate le varietà con la sola la caratteristica di produrre grandi quantità di uva e di essere particolarmente rustiche nei confronti del clima.

Alle soglie del rinnovamento post-fillosserico, però, si manifesta in tutta la sua gravità il problema del ritardo delle tecniche di vinificazione, ancora non adatte a valorizzare le caratteristiche qualitative delle viti francesi come chardonnay o merlot. Nel primo Novecento, la coltivazione di viti straniere in Italia è ancora un fenomeno sporadico e pionieristico, portato avanti da alcuni imprenditori più avanzati in Trentino-Alto Adige (allora chiamato Tirolo meridionale) e in Friuli-Venezia Giulia, dove c'erano alcuni centri di ricerca e formazione. La sostituzione delle vecchie varietà con vitigni di origine francese – e in parte tedesca – favorita dalla ricostruzione post-fillosserica diventa invece un fenomeno generalizzato che porta, in pochi anni, a un deciso cambio di scenario.

La diffusione di questi vitigni è inizialmente frenata dalla loro produttività, molto inferiore rispetto a quella delle varietà autoctone. Per valorizzarli, le operazioni enologiche vengono progressivamente modernizzate, segnando il passaggio da una produzione di vino destinata principalmente all'autoconsumo a una destinata al mercato. L'introduzione di uve francesi in Italia – lo abbiamo detto prima – punta alla produzione di vini simili a quei *Claret* bordolesi che tanto piacevano ai consumatori inglesi. Molto spesso, tuttavia, questi nuovi vitigni vengono utilizzati per migliorare il colore o l'aroma dati dalle varietà locali; il cabernet sauvignon è il vitigno più rappresentato. In Piemonte, nel 1860, il marchese

Leopoldo Incisa della Rocchetta lo introduce nella propria collezione ampelografica di Rocchetta Tanaro e lo descrive con le seguenti parole:

> «È un vitigno di robustezza ordinaria, ma vigoroso, che preferisce terreni più fertili rispetto al cabernet franc e produce una resa annuale uniforme. L'uva ha un sapore distintivo, ma un po' aspro. Il vitigno è tra i più pregevoli e merita di essere propagato».

La metafora del paesaggio vitivinicolo: tra realtà e immaginario

In definitiva, la grande storia della viticoltura italiana è il risultato di tante piccole storie, ancora da raccontare: un arcipelago di comunità, risonanze, isole culturali e borghi lontani che si sono formati attraverso la stratificazione di esperienze e popoli attorno a luoghi di grande transito e commercio. A queste popolazioni, che hanno conservato intatti gli immutabili riti della potatura, della vendemmia e della vinificazione, dobbiamo le più antiche tradizioni enologiche d'Italia.

Il paesaggio vitivinicolo diventerà sempre più il vettore essenziale per la conoscenza dei vigneti e dei vini di un territorio e, quindi, la risorsa più importante per tutte le strategie enoculturali. Il potenziale metaforico che possiede un vigneto è molto forte. Questo trasferimento di sensazioni dal paesaggio concreto all'immaginario è un procedimento comune operato da un vigneto. Il vigneto è, innanzitutto, una metafora di grande equilibrio per l'immagine che affiora da una natura antropizzata, di un'armonia tra uomo e pianta, una sorta di complicità. Ma è anche un'eloquente metafora del dinamismo. Il paesaggio è portatore di entusiasmo, ma allo stesso tempo di rigore e stabilità che confortano e stimolano il consumatore. Attenzione, però, alle implicazioni negative di un paesaggio poco rispettato o di un vigneto mal tenuto, che possono esprimere un conflitto irrisolto tra modernità e natura o essere portatori di un messaggio negativo.

È da sottolineare che gli elementi naturali del paesaggio acquistano significato solo in relazione ai contesti storici e culturali. Il paesaggio è quindi uno spazio costruito del quale il lavoro dell'uomo diventa parte integrante. Non è però solo ciò che appare all'occhio: i fenomeni territoriali sono governati da strutture latenti che devono essere indagate con strumenti adeguati.

Il paesaggio deve testimoniare l'autenticità di un vino. Cosa significa? La parola "autentico" deriva dal greco *authentikos*, che significa "agire di propria autorità" (*autos* = sé stesso e *hentes* = essere). Indica dunque essenza e reale interiorità, non apparenza. In relazione al vino, autentico significa quindi genuino, espressione di un territorio ben definito e difeso.
Astenendoci per un istante dall'immagine della natura e del paesaggio nel proprio evolversi secondo il gusto estetico occidentale, possiamo identificare il paesaggio viticolo intorno a un nucleo abitato, da un punto di vista eminentemente antropologico, come un giardino eutopico (*buon luogo*), ricorrendo al mito per fondare su di esso la memoria e l'identità dei suoi abitanti. Troviamo la descrizione di un giardino-vigneto nell'*Iliade* (XVIII, 561-566), dove si parla di alcune scene di vita campestre con le quali Efesto adorna lo scudo di Achille:

> «Vi pose anche una vigna, carica di grappoli maturi, una vigna d'oro, bellissima; neri erano i grappoli, e i tralci erano sorretti da pali d'argento. Intorno vi scavò un fosso di ciano, e all'esterno una siepe di stagno; e un unico sentiero conduceva al vigneto, per dove passavano i vendemmiatori ogni volta che raccoglievano l'uva».

È semplice riconoscere, in questa descrizione omerica del giardino dei Greci, tutte le caratteristiche del "giardino-vigneto" italiano: *aloe*, il campo piantato a vigna; *kepos*, il giardino alberato e ordinato; *horto*, la rigogliosa vigna circondata da un muro o da una siepe.

Un confine nascosto della viticoltura italiana: il sistema di allevamento

Nell'immaginario collettivo, la parola "vigneto" evoca un appezzamento di terra intensamente coltivato a vite, con filari ordinati e piante di vite sostenute da pali e fili di ferro. Se torniamo indietro di qualche decennio nella memoria paesaggistica, alla grande trasformazione che precede gli anni Sessanta e vede la scomparsa della mezzadria in molte aree dell'Italia centrale, ci rendiamo conto di quanto quell'immagine sia cambiata: il paesaggio dominante era al tempo rappresentato da coltivazioni promiscue, in cui la vite era "maritata" all'albero e all'olivo ed era accostata a colture erbacee.

Poche parole legate al vino possiedono il potere evocativo di *Enotria*. In passato, questa parola greca era erroneamente tradotta come "terra del vino" e identificava genericamente l'Italia. In realtà, *oinotron* significa "palo della vite" e dà origine al nome con cui i primi Greci sbarcati in Basilicata nell'VIII secolo a.C. chiamano quei luoghi, definendo *enotri* le popolazioni che utilizzavano aste per sostenere le proprie viti e che li avevano accolti pacificamente. Perché l'alberello con il palo colpisce così tanto la curiosità dei Greci da diventare il tratto distintivo di quei luoghi e di quel popolo? La spiegazione è molto semplice ed è, ancora oggi, sotto gli occhi di chi visita le isole greche come Santorini o l'entroterra turco di Smirne, dove la vite viene coltivata come tremila anni fa, con forme di allevamento striscianti.

Le forme a basso ceppo non erano le uniche presenti nelle aree più antiche della colonizzazione greca dell'Italia. Ciò è testimoniato da numerose fonti letterarie che descrivono, in epoca classica, la coltivazione della vite su sostegni vivi in Campania, similmente a quanto avveniva in modo più diffuso nell'Italia centrale e settentrionale. Significativa è poi la dicotomia paesaggistica tra le

alberate aversane, espressione della viticoltura etrusca, e – a poca distanza – l'alberello, segno della coltivazione della vite secondo modelli greci.
A partire dall'VIII secolo a.C., gli Etruschi di Capua e i Calcidesi di Cuma danno vita a una "frontiera nascosta" tra le due culture: quella della vite allevata ad alberello e quella su tutore vivo. Nelle campagne di Capua, infatti, dall'espansione etrusca fino a oggi, la vite si trova maritata al pioppo e, con la sua decisa impronta sul paesaggio agricolo, crea una profonda dicotomia rispetto alla vite allevata a ceppo basso nella Campania a sud di Napoli.
Il Greco di Tufo ha un'identità originale non assimilabile a nessuno degli altri vitigni chiamati greco. Potrà sembrare strano, ma questo vitigno, che tutti pensano figlio di una terra lontana, arrivato dal Peloponneso assieme a qualche pioniere avventuroso, è forse uno dei vitigni autoctoni più antichi d'Italia. Non è espressione di una cultura greca, bensì di una cultura etrusca. Ancora oggi, osservando come viene coltivato nella zona di Tufo o di Parete, nel Casertano, si trovano molte similitudini con le modalità di coltivazione utilizzate nella Pianura Padana dai paleoliguri: tre o quattro tralci sono attorcigliati in una sorta di treccia per collegare piante vicine. Questa tecnica si chiama *rumpotinus* e caratterizza i vitigni derivati dalla domesticazione di viti selvatiche con un elevato vigore e una vegetazione confusa e disordinata, ricca di femminelle. Della vite selvatica, inoltre, il Greco di Tufo conserva anche le foglie tonde, intere e molto particolari, e grappoli dalla forma singolare e irripetibile: un corpo principale, un'ala molto pronunciata e un evidente residuo stigmatico.
Geneticamente, il Greco di Tufo coincide con l'asprinio casertano. I due presentano forse qualche differenza fenotipica dovuta a processi di selezione antropica, ma il loro DNA è identico. Queste viti, alte fino a quindici metri, franche di piede e allevate con il

pioppo come tutore, in passato hanno sempre goduto di fama e notorietà poiché destinate alla produzione di un vino spumante ed esportate persino in Francia durante il periodo della fillossera.
Che il sistema di allevamento della vite su sostegni vivi sia un'eredità etrusca è dimostrato anche dalla presenza, nella lingua di questo popolo, della parola di radice indoeuropea *ataison*, che significa "girare su sé stesso", "attorcigliarsi", proprio come la vite maritata all'albero. Il termine "attorcigliarsi" è inoltre associato a vitigni provenienti da aree culturali molto lontane, come l'Asia Minore e la Sicilia orientale. Qualche esempio? Il nome del biblino, uno dei primi vitigni acclimatati nella Magna Grecia, significa letteralmente "vite attorcigliata" o "vite che si aggrappa"; "aglianico" ha come sinonimo *anadendrite.*
Gli agronomi greci del IV secolo descrivono i vigneti dell'Italia meridionale come territori costituiti da viti che si sostengono su tutori vivi, come pioppi e salici. Ciò dimostra che la coltivazione della vite su alberi non è esclusiva delle popolazioni dell'Italia centrale e della valle del Po, ma è presente anche in territori molto distanti. Se in alcune aree la diffusione delle due forme di allevamento si verifica con una precisa demarcazione, in altre, sotto la spinta della colonizzazione greca, i sistemi a basso ceppo danno origine a forme più espanse, come avviene nelle zone di produzione del Cecubo. Qui, ai confini settentrionali della Campania, il sistema di allevamento iniziale con pioppi come tutori vivi viene sostituito, in età imperiale, con il sistema a basso ceppo. Quest'ultimo riproduce modelli viticoli greci ai quali la cultura latina faceva riferimento, anche per l'impiego di schiavi provenienti dalla Grecia.
Anche se oggi è presente solo in vecchi vigneti marginali, l'alberello sostenuto da un paletto è un'iconema dal carico semiologico insostituibile, espressione di scelte compiute lungo un percorso di

secoli. Rappresenta un pertugio, una finestra attraverso la quale ci relazioniamo con un territorio inteso come spazio organizzato, come luogo in cui le componenti strutturali si fondono con quelle di forma e carattere. L'alberello, nella sua specificità, rappresenta quindi un oggetto di cultura, sia pure nel suo comporsi con la natura; almeno finché la cultura sarà il risultato di un processo lungo il quale una società trova la propria identità.
La maggior parte di questi iconemi è scomparsa. Quelli sopravvissuti sono ormai residualità, museali o marginalizzate, fuori luogo nei moderni contesti produttivi della viticoltura italiana. La versione odierna di questa disciplina è il frutto di un cammino di generazioni di viticoltori nel quale il susseguirsi di eventi ha dato origine a realtà così complesse da rendere impossibile riconoscere le singole fasi evolutive. L'alternarsi di espansioni e contrazioni – risultato di crisi socio-economiche, di editti ora favorevoli ora sfavorevoli alla coltivazione della vite, di cambiamenti climatici e di gravi malattie – ha portato alla semplificazione dei modelli viticoli (vitigni, forme di allevamento, tecniche colturali): la necessità di un'efficace meccanizzazione distingue la viticoltura competitiva su scala globale da quella destinata a scomparire per emarginazione economica.

La metafora del viaggio e la narrazione della viticoltura medievale

Il viaggio è una metafora della vita. Implica il superamento di difficoltà, prove di conoscenza, ricerca del nuovo, fuga dalla realtà. È uno strumento narrativo fondamentale sin dai *nostoi*, i viaggi di ritorno dalla guerra di Troia degli eroi greci nell'epica omerica. Lo *storytelling* è il format del percorso biografico dei personaggi legati al mondo del vino, che si sviluppa attraverso storie, saghe e racconti collocati in una dimensione temporale (un contesto storico) e in una dimensione spaziale (i luoghi del mito). Il viaggio – spesso metaforico, sentimentale, interiore – è una ricerca dell'eterna Atlantide, che vive nella leggenda e richiama qualcosa al di là della realtà.

È l'avventura stessa a generare la propria storia. La parola *aventure* nasce nel contesto gallo-romano e indica "le cose che ci vengono incontro in modo inaspettato". I viaggi medievali lungo i sentieri della fede sono avventure, così come è un'avventura la storia dell'Italia enologica lungo le vie consolari o dei pellegrinaggi. Persino la scienza è un'avventura.

Dopo la caduta dell'Impero Romano, la diffusione della vite avviene – come spesso nella sua storia – per merito di un dio e di una religione, e il vino diventa un soggetto sociale della civiltà europea. All'inizio, la coltivazione della vite è un fenomeno iterativo e locale, dettato dalla necessità di avere vino per il consumo alimentare nelle piccole comunità agricole. La crescente presenza di piccoli cenobi benedettini trasforma poi la viticoltura in un fenomeno collettivo fondamentale per l'economia di intere regioni, dove la vite assume il ruolo di coltura guida e i suoi prodotti non sono più destinati solo al consumo locale, ma anche all'esportazione. Queste relazioni tra le numerose comunità religiose, insediate soprattutto lungo le grandi vie di pellegrinaggio cristiano in un'Europa ancora senza confini, sono alla base non solo della diffusione di quei vitigni già presenti ai confini dell'Impero Romano, ma anche di un nuovo ciclo di incroci, più consapevole rispetto al passato nella selezione dei vitigni parentali. Pinot e traminer sono assoluti protagonisti, soprattutto nelle aree in cui la presenza monastica benedettina è più forte.

Tale aspetto della storia sociale del vino segna una profonda faglia tra l'origine dei vitigni dell'Europa continentale e delle regioni attorno alle Alpi, attraversate dalle grandi vie di collegamento tra sud e nord, e quelli delle zone affacciate sul Mediterraneo, dove il processo di domesticazione locale della vite risulta meno determinante rispetto al contributo culturale e genetico dell'esperienza coloniale greca.

Lo sviluppo delle vie della fede inizia nell'XI secolo nell'ambito di un progetto di trasformazione delle strutture economiche e sociali conseguente ai profondi cambiamenti dell'agricoltura altomedievale. In particolare, i collegamenti con le città marinare di Pisa e Venezia consentono l'accesso a merci provenienti dal Levante, soprattutto tessuti e spezie.

La "via di monasteri" – il tratto della Francigena che, dalla Pianura Padana, attraversa l'Appennino per raggiungere la Toscana e Roma, che ha in Bobbio il proprio punto focale e nel cui monastero di San Colombano compare per la prima volta la parola "Europa" – segue sostanzialmente il tracciato della strada romana e collega una fitta rete di monasteri e ospedali. È il cuore del *Regnum Langobardorum*, il *limes* nascosto con i territori bizantini situati più a est, lungo la Via Emilia.

A partire dal IX secolo, soprattutto nella Pianura Padana, alla viticoltura ecclesiastica dell'epoca carolingia si affianca una viticoltura aristocratica, nata per imitazione, ma che presto diventa una fonte di reddito. La diffusione del vigneto avviene principalmente nelle aree suburbane delle città, irradiandosi da esse, ma sempre a breve distanza dalle abitazioni per evitare i frequenti furti d'uva. Intorno all'anno 1000 si sviluppa la viticoltura borghese come conseguenza della crescita di una classe mercantile e artigianale con capitali da investire nelle campagne, specie nella produzione di vino, spinta dal forte aumento demografico delle città europee. Per antica consuetudine greca e latina, l'uva deve essere pigiata direttamente nel vigneto (*nullus habeat vel faciat mustum qui non habeat propriam vineam*). Per questo, lungo le strade si trovano sia le cantine sia i luoghi di smercio, anche se questo privilegio è riservato ai conventi. Le taverne hanno l'obbligo di essere situate entro le due miglia dalle città, questo per agevolare la riscossione del dazio comunale.

Le antiche strade non sono solo un filo conduttore tra le aree vitivinicole italiane, ma anche un elemento strutturante del paesaggio. In particolare, la centuriazione di quelle romane disegna lo spazio rurale secondo forme ortogonali: ogni centuria, composta da tre lotti, viene assegnata ai veterani legionari. Questo schema, tracciato dal *decumanus maximus* e dal *cardus maximus*, è ancora visibile nella viabilità interpoderale di alcune zone della Romagna, lungo la Via Emilia.

La viticoltura medievale italiana fotografata dalla tavola Peutingeriana

La cartografia, invenzione che precede la scrittura, ha l'intento di mappare l'ignoto e sovrapporvi una struttura. I cartografi dell'alto Medioevo usano le mappe come le pagine di un libro, per offrire una visione teologica della Terra agendo sull'immaginario di chi le consulta.

Questa rappresentazione dell'*eucomene*, una vera e propria forma di *storytelling ante litteram* del mondo conosciuto all'epoca, preserva la toponomastica romana fino al XV secolo. Si tratta di uno strumento cartografico originale, molto diverso dai mappamondi, in uso dall'alto Medioevo fino al Quattrocento e costituito da mappe regionali e carte nautiche o portolani. Come nella tradizione romana degli *itineraria* da cui deriva, la sua caratteristica principale è quella di tracciare una rete di linee, non di rappresentare una superficie continua.

La più illustre di queste carte è la Tavola Peutingeriana, una copia del XIII secolo di un originale tardo-romano risalente alla seconda metà del IV secolo, realizzato sotto l'auspicio di Teodosio. Disegnata da Castorius, la mappa prende il nome da Konrad Peutinger, il funzionario della città di Augusta che la ricevette nel 1507 dall'umanista e bibliotecario Conrad Celtes. La mappa si presenta come un rotolo lungo 6,80 metri e alto 0,34, e illustra il mondo allora conosciuto, dalla Britannia all'India. Disposta in fasce orizzontali, separate dal Mediterraneo e dall'Adriatico, riporta l'indicazione di montagne, fiumi e città, queste ultime rappresentate da gruppi di case. Piccole vignette riportano le distanze fra i vari elementi, calcolate in base ai giorni necessari per attraversarle. Sono inoltre menzionati anche i principali ostacoli e i modi per evitarli. Il tracciato topografico nel suo insieme è un'unità di comunicazione: gli elementi figurativi della carta sono necessari alla comprensione del messaggio che essa veicola. Il suo intento comunicativo è quello di raccontare, attraverso le vignette, la storia di città, conventi, chiese e castelli. Su queste tracce, aiutandoci con le testimonianze dei georgici latini e dei cronisti rinascimentali, troviamo anche il racconto di vigneti, vitigni e vini.

La Tavola è la colonna vertebrale di una fitta rete viaria europea. Un itinerario poteva essere preferito a un altro a seconda della stagione, dei conflitti in corso, delle condizioni politiche o per evitare zone infestate da briganti.

Nel tratto che interessa la penisola italiana, in corrispondenza delle diverse località, è possibile stabilire – sulla base di fonti latine (Columella, Plinio, Virgilio), medievali e rinascimentali – i vini prodotti e i vitigni utilizzati. È il primo grande tracciato attraverso l'Europa e ridefinisce il concetto di "strada", distinguendolo da quello di "via". quest'ultima, infatti, indica la modalità con la quale si procede, mentre la strada comprende l'insieme dei luoghi da attraversare per spostarsi da un punto all'altro. È anche possibile parlare di "vigneti di strada", intendendo l'insieme delle interrelazioni antropiche, politiche e sociali che hanno segnato i connotati viticoli lungo le strade di grande percorrenza. Spesso accade che queste coincidano con città di origine o romana (*oppidum, municipium, castrum*) oppure medievale, distinzione che ci permette di dedurre se, nella conformazione di questi vigneti e nella selezione dei vitigni coltivati, gli *hospita* religiosi abbiano avuto un ruolo determinante.

Gli attuali vigneti di Pavia, Piacenza, Lucca, San Gimignano e Viterbo sono l'eredità di quelli installati lungo la Via Francigena, così come quelli di Ascoli Piceno sulla Via Romea. Queste località, tra le più importanti tappe per i pellegrini diretti a Roma, conservano ancora oggi le tracce di quell'epoca nella posizione topografica originaria e nelle varietà coltivate.

Nel 1305, il bolognese Pier de' Crescenzi è il primo dopo i georgici latini a descrivere, nel suo *Liber ruralium commodorum*, circa quaranta vitigni e le relative collocazioni in diverse località dell'Italia settentrionale, tra Piemonte e Romagna. Fornisce così una precisa indicazione dei luoghi in cui queste varietà vengono coltivate e dove i vini da esse prodotti risultano celebri per la propria qualità. Rispetto ai riscontri dei georgici latini e dei neo-georgici rinascimentali, l'opera di Crescenzi offre una descrizione molto accurata delle caratteristiche morfologiche e produttive dei vitigni. Dopo

di lui, nei primi mesi del 1536, il bottigliere di Papa Paolo III Sante Lancerio scrive una guida vini *ante litteram*, descrivendo con cura e dettagli gastronomici a corredo i vini che incontra lungo le strade consolari romane. Percorre infatti lunghi tratti della Via Cassia e della Via Aurelia, da Roma a Nizza, della Via Emilia e della Via Flaminia, da Parma ad Ancona.
Lungo la Via Francigena e nelle sue intersezioni con le rotte del sale e dell'olio nasce la viticoltura medievale. I suoi modelli strutturali – caratterizzati da viticoltura promiscua, alberate e specifici vitigni – rimarranno fino all'arrivo delle malattie americane della vite.

La metafora dell'angoscia

Il sentimento dell'angoscia ha segnato gli ultimi duemila anni della nostra storia. La prima "età dell'angoscia" coincide con la caduta dell'Impero Romano, che fa mancare le certezze del diritto e la protezione dalle invasioni barbariche. Il Cristianesimo rappresenta allora l'unico rifugio dalla paura, e la necessità del vino per la celebrazione della messa diffonde la coltivazione della vite in tutta Europa attraverso il monachesimo benedettino.

Con l'intervento di Probo per cancellare l'editto di Domiziano, la vite torna a popolare i villaggi europei. Dai vitigni portati dalla Pannonia, antica provincia dell'Impero Romano che comprendeva alcune regioni delle attuali Austria, Ungheria, Slovenia e Croazia, nascono varietà in gran parte coltivate ancora oggi.

A cavallo del primo millennio si diffonde una delle prime *fake news* della storia: la fine del mondo è imminente. In Europa, il desiderio di espiare i peccati spinge i cristiani alla conquista del Santo Sepolcro attraverso le Crociate. Nascono i movimenti pauperistici francescani e i pellegrinaggi lungo le vie della fede che conducono a Roma diventano un fenomeno globale che sarà alla base dello sviluppo dei grandi centri della viticoltura europea lungo la via Francigena: Champagne, Alsazia, Borgogna, San Gimignano, per citarne alcuni.

Nell'iconografia medievale, la via Francigena è rappresentata anche come un labirinto, per esprimere il disagio e il senso di smarrimento che si prova nell'affrontare un viaggio nell'incertezza di poterlo portare a compimento. Il labirinto è infatti raffigurato, per esempio, su una colonna della cattedrale di Lucca e nella chiesa di San Pietro a Pontremoli, risalente al XII secolo. Il labirinto è un unico lungo, un tortuoso percorso da seguire che si rivela solo attraversandolo. È come una rete in cui ogni punto è connesso con un altro e che costringe il viandante a correggere continuamente la propria rotta, avanzando per tentativi ed errori. Il viaggio intrapreso da chi accetta di entrare in un labirinto e di attraversarlo con il rischio di perdersi, cercando un'uscita, è un'esperienza iniziatica di profondo significato religioso. Il vagare del viaggiatore, al tempo stesso stupefatto e confuso, pieno di dubbi e di svolte inattese, è una metafora della vita, nella quale giungiamo alla meta solo dopo un lungo percorso che porta alla "fine del mondo". Il labirinto è una struttura archetipica che riflette il modo di pensare dell'uomo, una sorta di adattamento all'incertezza del futuro. Ma il suo intrico è anche il mondo del peccato, dal quale il cristiano deve evadere attraverso la penitenza del pellegrinaggio.

Durante la IV Crociata, in un piccolo porto del Peloponneso chiamato Monemvasia, Venezia scopre un vino ancora sconosciuto in Occidente: quella Malvasia che farà da volano alle sue fortune commerciali durante la Piccola Glaciazione. Questa crisi climatica, che interessa circa quattro secoli tra Medioevo e Rinascimento, segna drammaticamente il destino dell'Europa con pestilenze, carestie e guerre di religione, ed è al centro della terza età dell'angoscia.

Anche nel mondo contemporaneo sono presenti molte paure che generano angoscia: pensiamo ad esempio al COVID-19, che ci ha messi di fronte alla transitorietà della vita e alla sua imprevedibilità, o agli effetti del cambiamento climatico e alla crescente scarsità di risorse. Se oggi manca il conforto della religione che vigeva in passato, torna il sentimento laico che vede nella Madre Terra il nostro rifugio. I movimenti ecologisti e il desiderio di sostenibilità hanno questa origine ancestrale. Tutelare la natura significa dare continuità al nostro futuro. Gli ideali di sostenibilità ambientale, centrali nel reagire agli effetti del cambiamento climatico, impongono un ritorno all'ecologia e a una viticoltura di basso impatto, ad esempio tramite la selezione di vitigni resistenti, di nuovi portainnesti più efficienti, di tecniche colturali con meno chimica. Richiede inoltre un minore consumo di combustibili fossili e di puntare sulla produzione di energie rinnovabili in azienda, secondo i principi dell'economia circolare.

La Piccola Glaciazione medievale: una lezione per il futuro della viticoltura europea

I secoli attorno all'anno Mille regalano all'Europa un clima caldo che spinge la viticoltura fino a latitudini molto elevate, addirittura fino alla Scozia, e ad altitudini altrettanto insolite nelle Alpi, attorno ai 1200 metri sul livello del mare. In questa fase, nota anche come *optimum* climatico, la viticoltura – accompagnata dall'allevamento dell'ulivo – si espande anche all'interno delle vallate alpine e occupa intere regioni dell'Europa continentale grazie all'opera delle istituzioni monastiche. Improvvisamente, però, tutto cambia.

L'iconografia medievale, soprattutto quella dell'Europa settentrionale, utilizza frequentemente temi che richiamano le condizioni di vita delle popolazioni rurali dell'epoca. Basta osservare i colori lividi dei dipinti di Bruegel il Vecchio (come *Cacciatori nella neve* del 1565) o le numerose scene popolate di scheletri dei cimiteri o dei luoghi sacri, le cosiddette danze macabre o *imago mortis*, per comprendere quale fosse il clima – non solo spirituale – di quei tempi. L'inclemenza del periodo è testimoniata anche da una figura dipinta da Ambrogio Lorenzetti a Siena nel 1338: un uomo completamente intabarrato su uno sfondo grigio, mentre una neve molto fitta cade tutto intorno e si deposita tra le pieghe del suo mantello. Questo modo di rappresentare i fenomeni naturali è una vera novità per il XIV secolo, e solo l'eccezionalità degli eventi meteorologici di quegli anni poteva ispirare il pittore del famoso *Allegoria del buono e del cattivo governo* a introdurla.

Un preciso riferimento alle condizioni climatiche delle regioni alpine verso la fine del 1300 si trova nel *Ciclo dei mesi* della Torre Aquila presso il Castello del Buonconsiglio a Trento. L'affresco rappresenta una scena di vendemmia nella quale si notano alcuni dettagli insoliti, ma molto efficaci nel rivelare la viticoltura di quel periodo: il sistema di allevamento ad alberello, più indicato delle pergole perché mantiene l'uva vicina al terreno sfruttandone l'effetto mitigante; il torchio coperto, per proteggere il mosto da pioggia e altre intemperie; l'abbigliamento dei contadini, molto pesante e adatto ad autunni freddi.

Cosa accade in Europa in questo momento storico? Verso la fine del Trecento, gelate, grandinate, piogge violente protratte per settimane e anni di siccità causano gravi carestie, creando il terreno ideale per il dilagare di pandemie come la Peste Nera. Quest'ultima giunge da Costantinopoli con le navi dei mercanti genovesi ed è descritta da Giovanni Boccaccio nel *Decameron*. L'epoca di Shakespeare, il Seicento, coincide con un periodo di incertezza e conflitti come le guerre di religione, la Guerra dei Trent'anni e quella di Successione Spagnola, esacerbato dallo stato di disagio delle popolazioni europee per le prolungate carestia e fame. Nel gennaio del 1709, un fenomeno anomalo battezzato dagli storici "Piccola Era Glaciale" distrugge in una notte l'intera viticoltura europea e si conclude solo attorno al 1850, con la *Irish potato famine*. È l'inizio della fine del feudalesimo. La localizzazione dei vigneti in Europa subisce un cambiamento drastico, scomparendo dall'Inghilterra e da tutte le valli interne delle Alpi.

Il momento viene reso ancora più critico dalla cosiddetta "guerra dei noli", che coincide con il cambiamento climatico e rende poco conveniente la produzione di vini a basso grado. Le nuove tariffe applicate al trasporto del vino, infatti, vengono ora calcolate sul valore del vino e non più sul numero di barili trasportati. Questo porta allo spopolamento di molte aree alpine e appenniniche e alla conseguente perdita di molti vitigni, frutto di processi di selezione risalenti a centinaia di anni prima.

La glaciazione, con la drastica riduzione dei cereali, del foraggio, della frutta e dell'uva, rappresenta una catastrofe per l'agricoltura. La produzione annua di vino è un eccellente indicatore della successione cronologica degli eventi climatici che possiamo consultare grazie all'abitudine, soprattutto da parte degli ordini monastici, di annotare la data di inizio della vendemmia, i prezzi correnti del vino e la qualità delle uve. Già dal XIV secolo abbiamo dati relativi ai vigneti della Borgogna e della Germania meridionale.

Come reagiscono le aree viticole europee al rapido raffreddamento del clima? La prima percezione del cambiamento coincide con la riduzione della produzione d'uva, a causa delle primavere fredde che ostacolano l'allegagione, e del titolo zuccherino per gli autunni piovosi, che rende difficile conservare il vino nel corso dell'anno successivo. A volte, le minime termiche invernali causano la morte delle viti. Le strategie di adattamento variano a seconda del Paese viticolo e riguardano le varietà coltivate, le forme di allevamento e l'utilizzo delle risorse energetiche dei terroir, alla ricerca di condizioni di esposizione dei vigneti che possano favorire la maturazione.

I primi interventi riguardano solitamente la forma di allevamento, con la quale si cerca di avvicinare il più possibile il grappolo al suolo affinché possa sfruttarne il calore irradiato. Vengono abbandonate le forme di allevamento su tutori vivi come le alberate, espressione della viticoltura primigenia, e si adottano sistemi di allevamento a basso ceppo (ad alberello) o a bassa pergola, detta "a camera". La viticoltura viene spostata in posizioni collinari, precedentemente destinate al pascolo per la loro scarsa fertilità. Questa scelta favorisce soprattutto le grandi proprietà nobiliari ed ecclesiastiche, che dispongono di vaste aree di terreno, le quali diventano così in grado di controllare l'intero mercato dei vini di qualità. Solo i vini del Mediterraneo orientale, commercializzati da Venezia, riescono a contrastare questo monopolio che, per alcune regioni vitivinicole come la Borgogna, l'Alsazia o la Renania, si conclude solo con la caduta dell'*Ancien Régime.*

L'innovazione genetica alla base della risposta al cambiamento climatico

La prima reazione dei viticoltori europei al cambiamento climatico è quella di privilegiare le varietà a maturazione più precoce e a bacca bianca. Ne dà testimonianza esemplare l'andamento della produzione di uva per ettaro tra il 1500 e il 1800 attorno al Lago di Caldaro, nel Tirolo meridionale, registrato dal Capitolo del Duomo di Trento, proprietario di vasti vigneti in quella località. Emergono due tendenze contrastanti: la prima, coincidente con la fase più acuta della crisi climatica tra la fine del XV e l'inizio del XVI secolo, vede ridursi le superfici coltivate a schiava – varietà a bacca rossa – e crescere quelle di vitigni a bacca bianca, probabilmente di numerose varietà; la seconda, dopo il 1700, inverte la tendenza e riprende la produzione di schiava, uva molto generosa, riducendo quella di uva bianca.

Si rivela però determinante la scelta di nuove varietà, attuata sia selezionando i semenzali sviluppatisi da incroci spontanei e che mostrano particolari doti di precocità nella maturazione dell'uva, sia introducendo viti provenienti da altre aree, climaticamente meno favorite. È questo il caso della Champagne, che abbandona la varietà a bacca rossa come il pinot noir ed estende la coltivazione del *gouais*, giunto con le legioni di Probo e che, incrociandosi

con un pinot ancestrale, porta alla nascita dello chardonnay.
In Veneto, in particolare nel Trevigiano e soprattutto nelle fasi più avanzate del ciclo termico freddo, coincidenti con il Grande Gelo del 1709, si verifica l'abbandono delle varietà a maturazione tardiva, spesso le migliori. Con il ritorno a condizioni climatiche più favorevoli, vengono impiantati vitigni molto produttivi ma di modesta qualità per soddisfare la forte domanda di vino del mercato europeo. Scompaiono così varietà come marzemino, bianchetta, malvasia e picolit e prendono piede vitigni più insipidi e acidi come verdiso, rabbiosa (raboso) e dall'occhio. In ogni caso, specialmente in Francia e Germania, la qualità dei vini rimane molto bassa nonostante gli adattamenti produttivi introdotti. Tali vini vengono destinati al consumo popolare: nobili e alto clero, insoddisfatti, cercarono altrove delle alternative.

La metafora di innovazione e concorrenza: Venezia

Una delle espressioni più celebri di Voltaire è *l'esprit des nations*, lo spirito delle nazioni. Esso definisce le caratteristiche strutturali, le eredità secolari e durature che conferiscono a un territorio un'identità inconfondibile, quasi un codice genetico. Nella formazione di questo spirito per le Tre Venezie, le regioni storiche di Venezia Euganea (Veneto), della Venezia Giulia (Friuli-Venezia Giulia) e della Venezia Tridentina (Trentino-Alto Adige), un contributo essenziale arriva dalla ricca tradizione vitivinicola.

Esiste un filo rosso che lega le innumerevoli tessere del mosaico che rappresenta i territori delle Venezie e che si può ben sintetizzare nella precoce e intensa attività mercantile legata al vino, testimoniata già dall'esperienza coloniale greca e dall'interesse degli Etruschi per il commercio negli empori di Adria e Spina. Nei prestiti linguistici, sopravvive ancora oggi la traccia del passaggio di molti termini legati alla viticoltura e all'enologia dal greco al latino attraverso l'etrusco, a testimonianza della base indigena della coltivazione della vite. Il contributo greco-orientale fu invece decisivo nel raffinamento della cultura del bere.

Non si può negare che un altro elemento unificante nella storia vitivinicola delle Venezie sia la presenza della Repubblica di Venezia in quest'area, dalle terre dell'Istria alla Vallagarina trentina. L'immagine del "leone di guerra", che campeggia sulle antiche porte d'ingresso delle città o negli affreschi dei palazzi più importanti, è un segno indelebile di una cultura che ha pervaso la vita delle popolazioni delle Venezie attraverso un linguaggio che, con un forte senso di appartenenza, le ha fatte sentire protagoniste dell'arte, della cultura e della civiltà europea dell'epoca.

Venezia insegna a questi popoli non tanto a produrre qualcosa - i veneziani erano noti nel Medioevo come gente che "*non arat, non*

seminat, non vindemiat" [non ara, non semina, non vendemmia], come riportato dall'anonimo autore di *Honorantie civitatis Papie*) - quanto ad aprirsi al commercio internazionale, distaccandoli dal bigottismo dei principi, dal vescovo del Tirolo e dalla decadenza del Ducato del Friuli. I mercanti veneziani erano eredi non solo degli Etruschi e dei Focesi che per primi avevano abitato le sponde dell'Adriatico, ma anche dei più scaltri mercanti levantini, dai quali avevano appreso le tecniche più raffinate del commercio.
La lunga storia del commercio del vino, nella quale i veneziani hanno un ruolo fondamentale, raggiunge la massima espressione con l'impatto della Piccola Era Glaciale sulla produzione vinicola in Europa tra la metà del 1300 fino al 1700 circa. Il commercio veneziano ha l'opportunità di controllare quasi tutti il mercato dei vini di qualità, dal Mediterraneo orientale fino alle ricche regioni dell'Atlantico settentrionale, favorendo la nascita e lo sviluppo in Italia di una viticoltura cosiddetta "di imitazione" per la produzione di vini dolci come Vinsanto e Malvasia, in sostituzione e integrazione della produzione mediterranea.
Si tratta di un'intuizione anticipatoria delle strategie commerciali che segnano il mercato del vino nei secoli successivi: le città marinare di Venezia e Genova commercializzeranno vini provenienti dal Mediterraneo orientale o dalle regioni del Midi francese e del Levante spagnolo utilizzando non il nome dei vitigni, ma quello – più o meno storpiato – del luogo di produzione, ad esempio Monemvasia, Santorini, Vernazza. Offriranno pochi vini, con caratteristiche organolettiche ben definite, facilmente riconoscibili e, grazie all'alto grado alcolico, trasportabili senza subire alterazioni nel sapore.
La concorrenza tra i commercianti di vino appartenenti alle diverse Repubbliche Marinare italiane affonda le proprie radici in un passato lontano. Riguarda da un lato vini dolci, aromatici e

alcolici, prodotti nel Mediterraneo orientale e commercializzati dai veneziani (Malvasia e Vinsanto) e, in misura inferiore, dai genovesi (Vernaccia ligure) sui mercati anglosassoni, dall'altro vini dolci ottenuti dalla viticoltura atlantica e tedesca con l'azione della muffa nobile, anch'essi destinati ai ricchi mercati del nord Europa, ma resi celebri soprattutto dai mercanti olandesi, ai quali era precluso l'accesso al Mediterraneo.

Nel ripercorrere la storia di una città, il pensiero va istintivamente alle origini, agli eventi che nel corso di anni e secoli ne hanno determinato la fama. Il passare del tempo ha però effetti inesorabili: i ricordi sfumano progressivamente e la ricostruzione dei fatti più remoti deve ricorrere a cliché, a suggestioni che, a poco a poco assumono i contorni del mito, della leggenda. Uno di questi miti identitari - forse il più noto - su cui si fondò la fortuna di Venezia è proprio quello legato al vino. Questa storia, tuttavia, non può essere raccontata solo attraverso la mera cronaca di eventi, battaglie e personaggi celebri, ma va trasformata in antropologia, restituendo a Venezia i suoi colori originari e facendo riferimento alla realtà fisica e sociale della vita quotidiana: è lì che nascono le idee. Il mito di Venezia è dunque impensabile al di fuori del quadro storico-sociale costituito dalla Repubblica, una città-stato caratterizzata dalla gestione assembleare del potere.

Per certi versi, queste riflessioni rendono i veneziani di allora nostri contemporanei, permettendoci di riconoscerci in loro; per altri, però, ce ne allontanano, facendoli piuttosto assomigliare alle società levantine da cui hanno recepito tanto. Sono due modi complementari per misurare somiglianze e differenze tra l'umanità odierna e quella del passato. Se questo ci avvicina agli antichi abitanti della Laguna, ci porta anche a scoprire la nostra alterità, quel fondo oscuro che ci rende stranieri a noi stessi, ma che – attraverso il mistero della storia di Venezia – ci permette di entrare nel paradigma dell'uomo moderno.

Venezia non è la più orientale delle grandi città occidentali, è una città orientale: basta osservare le cupole dorate della Basilica di San Marco al tramonto per rendersene conto. Anche le condizioni nelle quali Venezia prende forma durante l'Alto Medioevo favoriscono il consolidamento di mentalità e civiltà particolari, nate dallo speciale legame con Bisanzio e che fanno della città un punto di raccordo tra le grandi culture dell'epoca.

La Venezia altomedievale si distingue dall'Occidente dei feudi e dei castelli per la sua struttura economica eminentemente e precocemente mercantile. Sulle rotte commerciali, tuttavia, non viaggiano solo merci e denaro. Si diffondono soprattutto idee, che contribuiscono a formare attitudini e mentalità peculiari e che trovano una conferma originale nel senso dello Stato, nell'idea di bene comune vantaggioso per tutti, dai patrizi ai popolani. Venezia sa cogliere le opportunità offerte da alcuni aspetti particolari di quella fase del Medioevo: lo sviluppo delle città, dove si concentrava il consumo di vino, soprattutto nell'Europa settentrionale; la necessità, in un'economia di mercato, di approvvigionarsi di generi alimentari da territori lontani; la crescente domanda di una particolare tipologia di vino da parte delle emergenti élite sociali di mercanti e borghesia, stimolata dal suo valore di prestigio e dal conseguente alto costo che ne aumentano il valore simbolico.

Il tema della "rivoluzione commerciale", interpretata in funzione mediterranea, può essere collegato al mito italiano delle Repubbliche Marinare solo se messo in rapporto con la storia delle città di riferimento, intese come luoghi di scambio e commercio con i territori con cui interagiscono. Fin dal X secolo, Venezia mette in collegamento l'area adriatica, la regione balcanica, il mondo greco e l'Egitto con il ricco entroterra regionale, che la pone in costante e proficuo rapporto con il mondo tedesco. È il mercato globale, la rete dei suoi commerci a rendere Venezia una città tanto esotica – per modernità e contaminazione di costumi – quanto le sue controparti cinesi o indiane dove tutte le potenze commerciali d'Europa faranno scalo nei secoli successivi.

Tra il XIV e il XVI secolo, la società europea presta grande attenzione nel definire gli stili di vita dei diversi gruppi sociali. Le scelte alimentari e del vino in particolar modo sono strumenti privilegiati delle élite per marcare le differenze sociali. Il vino offre un elemento importante per codificare tali distinzioni: a ogni classe sociale corrisponde una tipologia di vino. Sono appannaggio delle classi abbienti i vini dolci di Cipro e i cosiddetti "vini greci", provenienti dai luoghi che coincidono con gli scali delle Crociate e occupati dai veneziani durante il trasporto dei soldati nella Quarta Crociata. Lo sono anche Grenache, Garnacha o Vernaccia, nomi che designano sia un vitigno sia un corrispondente vino di stile mediterraneo appartenente alla categoria dei vini dolci, liquorosi e ad alta gradazione alcolica.

In quel periodo si assiste a un cambiamento nella scelta del colore del vino. Come il pane destinato ai signori deve essere bianco e di frumento, mentre il pane nero di segale o farro è riservato ai contadini e ai servi, così il vino rosso o nero è destinato al "popolo lavoratore", mentre i vini bianchi e i *Claret* sono appannaggio di chi "vive nell'ozio". In questo hanno sicuramente un ruolo da un lato il cambiamento climatico causato dalla Piccola Glaciazione, che ostacola la maturazione delle uve a bacca nera, dall'altro l'arrivo in Europa di vini "greci" di colore bianco e ambrato.

Il primato dei vini dolci nel Medioevo è indiscusso e solitamente riferito a vini bianchi, imitazione di quelli provenienti dalla Grecia. L'innovazione rappresentata dalla bottiglia non si limita a prolungare la vita del vino e a facilitarne il trasporto, ma apre anche nuove opportunità commerciali, grazie alla possibilità di stoccaggio che consente non solo di controllare l'evoluzione del

vino nel tempo, ma anche di gestire l'offerta in modo da evitare cali di prezzo, specie quelli dovuti alla necessità di vendere il vino prematuramente per le precarie condizioni di stabilità e durata.
Le mutate condizioni economiche consentono a questi vini di diventare simboli di ricchezza e distinzione, e quindi di essere consumati dalle classi più abbienti. È curioso come la bottiglia diventi subito uno *status symbol*, e non solo per la sua praticità come contenitore. Circa un secolo prima, con lo sviluppo dell'arte vetraria a Murano, Venezia aveva già prodotto non solo pesanti bottiglie in vetro per il vino, ma anche una serie di accessori per servirlo, realizzati in vetro chiaro e di grande raffinatezza stilistica, che permettono per la prima volta al consumatore – abituato a contenitori di peltro, argento o stagno – di vedere in trasparenza il colore del vino che sta bevendo. Questa innovazione, che oggi può sembrare quasi banale, è ad esempio testimoniata nel dipinto del Veronese *Le nozze di Cana*, del 1563. I vini, soprattutto quelli dolci destinati alle tavole più importanti, vengono da quel momento serviti in caraffe trasparenti definite "alla moda veneziana", di magnifica fattura: diventa dunque importante la loro limpidezza e, di conseguenza, l'utilizzo di tecniche enologiche più raffinate rispetto al passato per la loro produzione.
Nel rigido cerimoniale che caratterizza i banchetti ufficiali della Dominante tra il XVI e il XVII secolo – dove le pietanze servite sono un *melting pot* di culture gastronomiche del Mediterraneo, con sapori agrodolci di ispirazione bizantina e carni o pesci esaltati dai profumi delle spezie più disparate – la "Corporazione del Tinello" incaricata dell'organizzazione del servizio apre ogni banchetto nello stile dei visir di Costantinopoli, ossia con pasticcini di pasta reale (marzapane) accompagnati da Marzemino dolce.
Per il re Enrico III di Francia, si apre la Sala del Maggior Consiglio e il menù dell'occasione comprende decine e decine di porta-

te. Il sovrano, frugale e inibito dalla paura dei veleni, non assaggia nulla ma non rinuncia a sorseggiare un bicchiere di Marzemino dolce. È questo l'omaggio che Venezia rende alle fonti della propria ricchezza: quei vini, quelle spezie, lo zucchero di Cipro, il sale e le migliaia di altre merci provenienti da tutto il Mediterraneo, che riempiono i suoi fondachi e testimoniano la sua forza e prosperità nei mercati del ricco Nord Europa.

Stiamo parlando della Venezia dei 17 teatri, tappa del Grand Tour, identificata nel Settecento come la capitale europea della frivolezza, di arte scenica senza pari, della musica barocca e dei costumi rilassati. È il contesto in cui Mozart immagina il *Don Giovanni* e in cui Emanuele Conegliano, *alias* Lorenzo Da Ponte, dopo l'abiura della religione ebraica, muove il proprio "doppio" di Leporello e Don Giovanni, coscienza e dissolutezza. Eclettico e amante della bella vita, assiduo frequentatore di salotti e sale da gioco, Da Ponte non può fare a meno di menzionare, nel libretto di *Don Giovanni*, "l'eccellente Marzemino", vino di moda tra l'aristocrazia veneziana insieme al Tokaj. Con ogni probabilità, Da Ponte ricorda il Marzemino che i conti Collalto, ambasciatori della Repubblica presso l'Impero, servono nel proprio palazzo di Vienna in *Platz am Hof* dove Mozart debutta, assieme alla sorella, il 9 ottobre 1762.

Il mito del Marzemino dolce rimane vivo fino alla fine dell'indipendenza della Repubblica, avvenuta il 17 ottobre 1797. Dopo quella data, ironicamente, ne diventa il simbolo non più del potere, bensì della decadenza.

Forse pochi sanno che Venezia perde il mercato dei vini dolci nel Nord Europa – a un passo dalla creazione dei "grandi vini" con il taglio di partite provenienti da località molto diverse – perché sottovaluta, alle soglie della "rivoluzione delle bevande", il futuro ruolo dello *"spirito di vino"* (l'alcol prodotto dalla distillazione del vino) nella creazione dei vini fortificati e dei mosti muti, di prezzo molto più conveniente rispetto a quelli importati dal Mediterraneo.

Venezia commette anche un altro errore: risponde ai vini fortificati – in particolare il Porto, che si basava su capitali inglesi per la produzione e il commercio – con la produzione del Vermuth, ispirato all'assenzio allora molto di moda sul mercato francese. Questo passaggio dalla Malvasia, da sempre identificata con il commercio veneziano, a una diversa tipologia di bevanda si rivela un'innovazione controproducente: i consumatori la percepiscono infatti come sinonimo di contraffazione del vino.

Per circa un secolo, il mercato dei vini fortificati è nelle mani degli olandesi, grazie alla loro grande potenza marittima. All'alba della "rivoluzione delle bevande" passa poi agli inglesi, che adottano una politica protezionistica e rimpiazzano i vini francesi (ad esempio i Moscati fortificati del Midi) con vini spagnoli e portoghesi. Anche gli inglesi praticano la mescolanza di prodotti alcolici di varia natura come regola fondamentale. Tuttavia, il rischio di essere accusati di sofisticazione viene scongiurato da due grandi innovazioni: le scoperte della chimica e della microbiologia, che consentono di produrre vini molto gradevoli e privi di difetti, e l'evoluzione del gusto dei consumatori, che si allontana dai *Claret* di Bordeaux per aprirsi a nuove esperienze, a bevande dagli aromi molto diversi da quelli del vino.

Italia e vino: un'idea di futuro

Questo mondo in continuo mutamento rende incerti i legami culturali e porta le persone a mettere in discussione le proprie radici, costringendole ad adottare regole altrui. Per evitare che la standardizzazione porti alla scomparsa delle culture originarie, è necessario inventare nuove forme culturali. È ciò che accade quando due culture diverse si incontrano: spesso, le cose più interessanti accadono ai margini.

L'Italia è il risultato di molteplici apporti culturali provenienti da ogni angolo d'Europa: tedeschi, francesi, austriaci e spagnoli, fra gli altri. Nel tempo, con l'espansione e la mescolanza delle culture originarie di diversi sistemi, la periferia è diventata il centro. È il frutto di apporti di vitigni e tecniche viticole dalle origini lontane e misteriose che, presi singolarmente, appaiono incompleti, ma che in questa viticoltura trovano una sintesi perfetta: una fusione di prodotti, stilemi e linguaggi che soddisfa il desiderio di eclettismo e sincretismo del consumatore moderno. È eclettico chi si muove con disinvoltura combinando diversi stili. È sincretico chi riesce ad armonizzarli, creando una sintesi seducente.

Se la principale risorsa della tradizione sta nella sua capacità di tutelare e tesaurizzare la diversità, è proprio nell'uniformità della trasmissione acritica del passato che le tradizioni muoiono. Per dare un nuovo futuro alla nostra viticoltura non dobbiamo affidarci né al tradizionalismo né alla sola innovazione tecnologica, ma partire da una corretta interpretazione della tradizione, focalizzando l'attenzione su un nodo cruciale della filiera vitivinicola: la comunicazione dei valori della nostra viticoltura.

La nascita della scienza enologica nel XIX secolo coincide con l'azione terapeutica del tecnico contro le numerose "malattie del vino" e, di conseguenza, contro i difetti gustativi che queste provocano. Così, la tipicità legata al territorio, una caratteristica distintiva del vino italiano, non è solo una questione di origine geografica, ma anche di fattori umani, di riconoscimento sociale e di appartenenza alla memoria.

Come possiamo allora restituire alla parola "tipicità" il significato originario, che rende un vino italiano quasi istintivamente identificabile dalle sue caratteristiche cromatiche o tipologiche, così come riconosciamo una persona dai tratti salienti del volto o un quadro dalla sua bellezza? Se associamo il significato di termini come "originalità", "identità", "naturalità", "tradizione" e "tipicità" (quest'ultima in senso weberiano), tutti valori che stanno alla base di un prodotto agroalimentare di qualità, alla percezione che il consumatore ha oggi dei vini DOC, ci accorgiamo che questo rapporto è solo semantico, non sostanziale. Le cause di questa frattura sono molteplici, ma ciò che gioca un ruolo decisivo nel fallimento della maggior parte delle denominazioni italiane è l'assenza di una cultura del territorio nei produttori, cultura che si sviluppa nel rapporto tra le caratteristiche dell'ambiente e quelle del vitigno.

La produzione del vino viene così interpretata come un esercizio demiurgico in cui la tecnologia assume un ruolo fondamentale, mentre l'uva, con le sue caratteristiche compositive, diventa solo una matrice da correggere in vari modi per ottenere vini adatti alle mutevoli esigenze del mercato. Alla base di un processo produttivo come quello del vino, così strettamente legato alla civiltà dei popoli mediterranei, deve invece esserci un principio etico. Forse il legislatore, cinquant'anni fa, sperava di dare un'anima alla legge sulle DOC, ma questo intento non è purtroppo emerso.

Se ci astraiamo dalla semplice espressione topografica e geografica di un territorio vitivinicolo e lo osserviamo per i significati profondi che emana, esso ci appare come una tavola, dove il *logos* (la

parola, il pensiero) si materializza in un vino. È una tavola come quella dei comandamenti offerta da Mosè al popolo ebraico, o come l'altare del sacrificio su cui si sancisce il patto tra uomo e divinità. Il territorio vitivinicolo è proprio una tavola di questo tipo: un luogo dove si realizza l'alleanza tra produttore e consumatore, nel rispetto delle regole della denominazione e delle sempre più fragili risorse ambientali.
Rispettare la terra significa chiederle ciò che essa può realmente darci. Questo atteggiamento trasforma il lavoro in un'alleanza con la terra. Il rispetto, però, presuppone innanzitutto una conoscenza approfondita di ciò che ne è oggetto: per esprimerlo con parole di Sant'Agostino,

> «Non si può amare ciò che non si conosce».

Gli studi dei terroir, condotti negli ultimi anni con approcci metodologici sempre più complessi nel tentativo di svelare il legame segreto tra le caratteristiche sensoriali di un vino e il suolo in cui cresce la vite, hanno certamente contribuito a rimuovere alcuni limiti nella comprensione dell'interazione tra vitigno e ambiente. Tali studi cercano di far prevalere gli argomenti tecnici su quelli giuridici delle delimitazioni territoriali, ma soprattutto evidenziano il ruolo sempre più rilevante del consumatore nelle scelte d'acquisto del vino, non solo per le sue qualità organolettiche, ma anche per i valori simbolici legati al luogo di produzione. La zonazione viticola diventa uno strumento capace di mediare tra le diverse emozioni suscitate da un paesaggio. Cerca un equilibrio tra cultura e natura al fine di offrire indicazioni su come salvare rappresentazioni simboliche ed esigenze ambientali, per raccordare istanze estetiche e fattori economici, per rispettare tensioni produttive e bisogni turistici.

Le origini e le radici, di per sé, non bastano. Sono una condizione necessaria, ma non sufficiente. Marc Bloch, uno dei *maître à penser* del Novecento, spiega questo concetto con un esempio: senza la ghianda, la quercia non esisterebbe, ma non tutte le ghiande diventano querce; se lo diventano, è solo perché in quel clima e in quel suolo la ghianda può esprimere il proprio potenziale. Questo potenziale, che alcuni vitigni esprimono solo in aree molto specifiche d'Italia, non è rappresentato solo dalle caratteristiche del suolo e del clima, ma soprattutto dal *savoir-faire* dei suoi viticoltori. Manca però ancora qualcosa di importante, ossia la consapevolezza che il vino, per i produttori italiani, è un bene comune come l'acqua, la scuola e l'ambiente. Ecco perché la viticoltura deve essere sostenuta da un patto sociale: non possiamo vivere bene se non ci sentiamo, prima di tutto, parte di un destino comune, se non maturiamo un maggiore spirito di corpo e un orgoglio di appartenenza a una terra così ricca di storia e risorse naturali che ci porti a superare localismi e gelosie.

Alla base di questa complessa fenomenologia vi è la polarizzazione della viticoltura italiana avvenuta negli ultimi anni per effetto della globalizzazione dei mercati internazionali. Si è passati da una realtà diffusa, rappresentata da una molteplicità di piccole denominazioni in cui i vitigni autoctoni erano l'elemento qualificante dei rispettivi vini, a una viticoltura che imita il modello francese, caratterizzato da cinque o sei denominazioni di fama internazionale che, da sole, rappresentano oltre il 60% delle esportazioni.

Questa opacità nelle scelte dei consumatori causa una crescente disaffezione specie tra i viticoltori più anziani, il cui reddito si è progressivamente eroso. Ciò porta al trasferimento delle autorizzazioni per gli impianti dalle regioni meridionali alla ricca viticoltura del nord-est dove, accanto a vitigni autoctoni consolidati come glera o corvina, il pinot grigio e le varietà destinate alla

produzione di spumanti sono in continua espansione.
Nello scenario economico e sociale descritto mancano inoltre studi universitari approfonditi su aspetti cruciali come il patrimonio viticolo autoctono, la ricerca viticola, i modelli di coltivazione adatti ai vitigni antichi, il controllo della produzione per ceppo, le dinamiche dei processi di maturazione e gli effetti del cambiamento climatico. Quest'ultimo, in particolare, potrebbe essere il contesto ideale per valorizzare le peculiarità compositive di alcuni vitigni (tannini, acidità, stabilità del colore) che in passato erano considerati inadatti alla produzione di qualità.

Affrontare il futuro conoscendo il passato

Se non conosciamo il passato, non possiamo prepararci per il futuro. Oggi andiamo troppo di fretta, ma se avessimo la pazienza di ripensare agli eventi che hanno plasmato la storia del vino europeo troveremmo spunti preziosi per pianificare meglio il domani.
Con la cosiddetta "rivoluzione delle bevande" di fine Seicento, il consumo di vino e distillati – così come la diffusione di bevande analcoliche di origine non europea come caffè, tè e cioccolato – subisce una profonda trasformazione che provoca una crisi in alcune delle regioni vinicole più celebri della Francia. In particolare, nel bordolese, i *Claret* destinati al mercato inglese vengono sostituiti dai vini più scuri di Cahors. Analogamente, la viticoltura si sviluppa in territori poco adatti alla produzione di vino di qualità, ma favorevoli alla distillazione (Cognac e Armagnac) anche grazie alla vicinanza a porti atlantici, strategici per l'esportazione nei paesi del nord Europa. La vera novità, però, è la nascita di vini come Champagne e Porto.
Quali furono le cause di questi cambiamenti? Molte le abbiamo già menzionate nei capitoli precedenti. Tra i fattori più significativi ricordiamo gli eventi climatici e demografici: la crisi climatica

del Medioevo che culmina con il Grande Gelo del 1709, l'arrivo della Peste Nera a Venezia e Genova nel 1348, le guerre di religione e le conseguenze sociali della Riforma Protestante. Anche se con impatto minore, contribuiscono fenomeni come l'arrivo in Europa delle malattie americane (fillossera, peronospora e oidio), i mutamenti nella distribuzione geografica della viticoltura dovuti all'aumento dei noli di trasporto del vino nel Medioevo, la scomparsa della viticoltura d'alta quota nelle Alpi, i cambiamenti economici e sociali alla base della conversione della viticoltura da promiscua a specializzata e viceversa, gli stimoli culturali ed economici provenienti da altri paesi, soprattutto nelle zone di confine, la circolazione di vini di altre aree vinicole, le mode alimentari.

Verso la fine del XVII secolo, l'industria del vino entra in una fase di innovazione permanente. Tre fattori contribuiscono in particolare: conoscenza, concorrenza e capitale, applicati soprattutto nella produzione di vini pregiati da invecchiamento e in quella di spumanti. La conoscenza si riferisce ai progressi nella chimica della fermentazione, all'identificazione dei lieviti come agenti fermentativi, all'uso dell'anidride solforosa come antisettico e conservante e alla produzione industriale di bottiglie e tappi di sughero. Il capitale è una condizione necessaria per chi produce e acquista vini da invecchiamento, sia per sopportare i tempi di immobilizzo in cantina, sia per coprire i costi di bottiglie e tappi prima di aver realizzato un profitto.

Nella seconda metà del XX secolo, l'enologo francese Émile Peynaud ispira un cambiamento di prospettiva, trasformando l'enologia in una pratica estetica finalizzata a ridurre l'effetto del caso sul gusto del vino. Il vino che gustiamo oggi è ben diverso da quello che si beveva anche solo pochi decenni fa. L'invenzione della tradizione vinicola e della sua ortodossia è un tratto della cultura popolare contemporanea che non ha fondamenti storici. Il rischio

più grande per questo importante comparto è quello di cadere vittima della propria retorica: una visione realistica dei comportamenti e delle tendenze di consumo è essenziale per consolidare i presupposti economici del settore.
A cavallo tra il XIX e il XX secolo, la consapevolezza che le condizioni di vita sono migliorate grazie ai risultati della ricerca scientifica contribuisce a diffondere l'idea che la scienza sia il motore del progresso umano, in grado di risolvere qualsiasi problema e garantire una prosperità inimmaginabile in passato. Questo ottimismo incentiva la nascita di collaborazioni tra il mondo accademico e l'industria, specialmente nei settori della chimica e delle applicazioni della fisica elettrica.
Ciò che i sociologi chiamano "spazio socio-culturale" è una dimensione virtuale, costruita dall'esperienza collettiva, che assume le caratteristiche di una carta topografica in cui sono indicati luoghi (della mente), traiettorie (percorsi ideali) e destinazioni (piacere). Come nell'esperienza kantiana, la nostra carta (topografica) mentale si identifica nelle variabili temporali e spaziali, che ci permettono di declinare la qualità in tre sfumature: anzianità, la qualità immanente dell'età; autenticità, la qualità certificata dalla tradizione; stabilità, la qualità acquisita per la durata.
Il declino di un'economia ha sempre a che fare con la sua incapacità di adattarsi alle trasformazioni del mercato, sia in termini di gusti dei consumatori che di metodi produttivi. Questo accade nel XVII secolo a Venezia, che si trova di fronte alla concorrenza di inglesi e olandesi, più abili nell'adattarsi al potere d'acquisto e alle preferenze della nuova clientela e che subentrano nel controllo dei porti prima veneziani nel Mediterraneo. A ciò si aggiunse un attaccamento alla tradizione che si traduce nel rifiuto dei nuovi vini "atlantici", richiesti invece dalla ricca borghesia europea.

Wine *storytelling* e giovani: una "enosofia" del bere responsabile

Una recente indagine HBSC (*Health Behaviour in School-aged Children*), uno studio internazionale sulla salute e il benessere degli adolescenti, ha rivelato che il 29% dei giovani attorno ai 15 anni consuma alcol almeno una volta a settimana. Raramente si tratta di vino: preferiscono in genere gli *alcopops*, miscele di succhi di frutta e superalcolici, più subdoli in termini di dipendenza. Bere è considerato *cool* e aiuta i giovani a superare le insicurezze tipiche della pubertà, diventando una sorta di rito di passaggio. Il vino viene consumato in modalità *binge drinking*, come espressione della trasgressione adolescenziale.

Servirebbe un corso di cultura e civiltà del vino, da inserire già nella scuola primaria, che coinvolga anche la formazione professionale degli operatori della ristorazione. L'attuazione di un progetto di tale portata necessita, naturalmente, di un piano formativo adeguato. Negli anni '90, in risposta ai crescenti problemi legati all'alcol tra i giovani e alla necessità di un intervento precoce nell'educazione alimentare, nasce in Italia l'*Osservatorio Permanente sui Giovani e l'Alcol* (OPGA): una comunità di ricercatori, esperti e comunicatori unisce forze e competenze con l'obiettivo di fornire un'informazione qualificata sul rapporto tra giovani e alcol, promuovendo un approccio antiproibizionista e orientato alla minimizzazione del danno. L'OPGA istituisce inoltre il "Laboratorio sui comportamenti e gli effetti del consumo di alcol tra i giovani" presso l'Università di Perugia, il quale collabora con i docenti dell'Ateneo in un progetto per prevenire il consumo a rischio di alcol. Queste iniziative, pur meritorie nelle intenzioni e valide come punto di partenza, non ottengono l'impatto diffuso e pervasivo che potrebbe invece sortire una legge *ad hoc*. Se dunque

l'Italia non si trova proprio all'anno zero in questo campo, per non disperdere il lavoro svolto finora deve mettere in atto provvedimenti più incisivi e coordinati.
Il poeta francese Charles Baudelaire scrive:

> «Il vino è simile all'uomo: non si saprà mai fino a che punto lo si possa stimare o disprezzare, amare o odiare, né di quali sublimi azioni o mostruose colpe sia capace».

Solo la cultura e la conoscenza possono fare la differenza.
All'inizio degli anni '80, il docente dell'Università di Montpellier Masquelier afferma che l'alcol del vino è come un albero che nasconde la foresta. Con questa metafora critica il cliché che identifica il vino con il suo principale componente, l'etanolo, oscurandone le proprietà salutari dovute a centinaia di altri composti presenti.
A quel tempo, l'opinione pubblica francese è schierata contro i pericoli dell'abuso di alcol, ritenuto dalla maggioranza dei medici – soprattutto gastroenterologi ed epatologi – responsabile di gravi dismetabolie. Anche i sociologi urbani vedono nel consumo precoce di vino un fenomeno con gravi ripercussioni sulla delinquenza e sull'abbandono scolastico.
La Francia ha familiarità col problema dell'alcolismo già dalla fine del XIX secolo, con la diffusione di distillati di cereali e vino che coincide con la rivoluzione industriale: intere generazioni di contadini malnutriti si trasformano in operai stanchi e frustrati, trovando nell'alcol l'unico conforto alla propria disperazione. Nascono in questo periodo i primi movimenti antialcolici, sostenuti anche dall'ideologia socialista fortemente contraria al consumo di alcol da parte del proletariato.
Purtroppo, l'assimilazione del vino all'alcol ha prevalso sull'idea

del vino come bevanda salutare fino ai giorni nostri. Solo l'eccezionale impatto mediatico del "paradosso francese" sul pubblico televisivo americano alla fine degli anni '80 ha riportato il vino – in particolare quello rosso – al centro della dieta moderna e dello stile definito "mediterraneo", noto per i suoi effetti benefici sull'incidenza delle malattie cardiovascolari. Questo stile di vita, e l'abitudine di bere vino a tavola in famiglia, hanno inoltre attenuato l'associazione tra vino e trasgressione, rendendo il fenomeno dell'alcolismo meno grave nelle popolazioni mediterranee rispetto a quelle del Nord Europa.

Introdotta in Francia nel 1991, la *loi Évin* vieta ogni forma di pubblicità a vini, superalcolici e tabacco. Nella sua *ratio*, la legge ha l'obiettivo di trasformare il vino da bevanda eccitante a strumento di cultura nazionale, partendo dalla formazione nella scuola primaria e secondaria. Vengono infatti stanziate risorse significative per organizzare mostre itineranti che istruiscono gli studenti francesi sulla produzione delle denominazioni più importanti, illustrandone i terroir, i vitigni, le tecniche di coltivazione, vinificazione e distillazione. Attraverso un approccio interattivo, le iniziative contribuiscono a far conoscere i vini anche attraverso il loro profumo, con dei diffusori aerosol che ne riproducono i profili aromatici. Il risultato? Una riduzione dell'impatto negativo dell'alcol tra i giovani e, al contempo, un rafforzamento del senso di appartenenza a un Paese che ha sempre fatto del vino un emblema della propria cultura. Di recente, una nuova proposta di legge presentata all'Assemblea Nazionale con il sostegno dell'associazione *Vin et Société* ha pianificato di sostituire questo ormai obsoleto strumento di comunicazione con l'istituzione di una rete televisiva dedicata esclusivamente al vino: alla diffusione della conoscenza del settore, alla discussione delle problematiche produttive, all'analisi delle tendenze di consumo, all'aggiornamento sulle sfide del mercato.

Raccontare il vino ai giovani

I valori, le emozioni dei giovani e il vino

Il *wine storytelling* per i giovani deve riflettere le differenze antropologiche tra le generazioni.

Baby Boomers: ormai "diversamente giovani", sono nati tra il 1946 e il 1964 e sono quindi figli del *boom* demografico del dopoguerra e del miracolo economico. Oggi rappresentano una sostanziosa componente della popolazione attiva, tale da determinare le sorti della nazione: basti pensare a Bill Gates, nato nel 1955, uno degli uomini più ricchi d'America, o a Donald Trump, nato nel 1946. Mostrano rispetto per la famiglia e la religione, sono ansiosi di guadagnare molto denaro e molto in fretta. L'unica vera differenza rispetto agli standard dei loro genitori è una maggiore sensibilità alla qualità della vita. La salute fisica viene subito dopo e implica la diffusione di jogging, aerobica, alimenti naturali e integratori vitaminici. La BMW è il loro *status symbol* preferito.

Generazione X: nati tra il 1965 e il 1980 e cresciuti in un mondo ossessionato – a livelli di paranoia collettiva – da valori come la pace, l'ambientalismo, il ritorno alla tradizione, l'ottimismo. Sono sovra-istruiti, sottoimpiegati, disillusi e sgomenti, lavativi nichilisti con un umorismo tagliente. Hanno attraversato la crisi energetica, l'AIDS e la recessione. Figli del divorzio, sono cresciuti davanti alla TV, ispirati dal "neo-tradizionalismo", ovvero un ritorno ai valori del passato, senza rinunciare ai progressi sociali, economici e tecnologici. Nel romanzo *Ologramma per il re*, il viaggio di un imprenditore fallito che cerca salvezza nel deserto diventa metafora di una generazione senza identità, come l'ha definita Douglas Coupland. Per sentirsi vivi, hanno bisogno di azioni, di esperienze individuali tangibili.

Millennial: nati tra il 1981 e il 2000, sono la generazione Erasmus, liberi dal servizio di leva obbligatorio. Istruiti, ma disperatamente disoccupati, emigrano in massa come i loro bisnonni. Sono abbastanza liberi dal passato da potersi permettere di rimpiangerlo, reinventarlo, contaminarlo: mettono insieme nostalgia del vecchio e *commodities* del nuovo, vitigni antichi e biodinamica, anfore e web. Sono un *ready-made* di passioni, ossessioni, credenze, tendenze, utopie e malinconie in continuo aggiornamento. Generazione social, interessati alla cultura, disincantati ma preoccupati per il futuro. Per loro, il cibo è sinonimo di identità e fonte di benessere: preferiscono prodotti freschi, se non biologici, non solo come esperienza organolettica, ma anche culturale e territoriale. Praticano il turismo enogastronomico.

Generazione Z: nati dopo il 2000, alla fine del vecchio millennio, sono definiti dalla lettera "i": iPad, iPod, iPhone. Hanno il digitale nel DNA e una straordinaria capacità di interagire con altre culture, anche perché coetanei degli immigrati di seconda generazione. Giorgio Ghiotti, millennial classe 1994, nel suo *Dio giocava a pallone* racconta storie di primi amori e prime trasgressioni dei *teenager* della Gen Z. Scrive, inoltre:

> «Quando vado nelle scuole a parlare di libri, vedo ragazzi affamati di emozioni, esattamente come lo eravamo noi alla loro età. Ascoltano, ma non vogliono indicazioni né sentieri già tracciati. La Generazione Z cammina con le proprie regole, e gli adulti possono solo adattarsi».

L'attualizzazione del mito del vino nei giovani: un percorso di conoscenza e conquista

Conoscenza e gioia

La bontà originaria: il vino è intrinsecamente buono. Le prime gocce date dal nonno, un rimedio contro il raffreddore, un mezzo per celebrare la caduta del primo dentino ne fanno inoltre una bevanda familiare.

La trasgressione: il vino è una bevanda degli adulti, diventa il simbolo dell'essere grandi. Prima è la birra, poi gli aperitivi (spumante, spritz), le occasioni, le gite scolastiche, le feste estive, i compleanni, la ricerca dello stordimento, le prime ubriacature.

Saggezza e passione

La morte di Dioniso: il suo mito è cosa per intellettuali, elitario. I giovani lo identificano con i genitori, con le loro conoscenze sui vini. La ricerca di vini veri, genuini (vini naturali e biodinamici) e anti-snob rappresenta il taglio del cordone ombelicale.

La relazione: diventare grandi significa uscire dal gruppo, come insegna il *Jack Frusciante* del romanzo di Enrico Brizzi. Uscire dal gruppo, però, vuol dire conquistare un nuovo livello di capacità relazionale: si scelgono nuove amicizie, si approfondisce la comunicazione. Il vino cessa di essere un elemento trasgressivo per diventare un fattore disinibente, per ridurre le distanze. È tuttavia necessario conoscere i diversi vini, come si fa per i cibi, quasi per difendersi dalla parte cattiva del vino, quella dell'alcol. Conoscere è una forma di controllo e un modo per esprimere sé stessi, il segno della transizione verso il mondo adulto.

Disinibizione e controllo

La cura come ritorno alle origini: la ricerca della verità, rendere visibile ciò che è invisibile attraverso l'esperienza e la conoscenza. Fuori dall'adolescenza, il vino diventa un modo di espressione, di prendersi cura di sé stessi e delle proprie relazioni. Diventa un mezzo espressivo per collegare l'attrazione idealizzata dell'infanzia all'impatto deludente dell'adolescenza e alla consapevolezza dell'essere adulti, ma anche a un prodotto sano legato alla terra, alla sua provenienza.

Territorio e tradizione

La terra è l'etichetta: il vino buono è quello delle proprie origini, non quello australiano o californiano che non ha tradizioni e non sa accompagnare i cibi. È il vino di territori ben delimitati, con caratteristiche territoriali specifiche che lo studio dell'origine geologica dei *terroir* esalta. Tutti i vini identificati da un'origine geografica sono buoni. Qui entrano in gioco la fascia di prezzo e l'etichetta (diversi *terroir*, diversa qualità dei vini in base anche alla loro fama, ad esempio Sassicaia o Tignanello) e si apre lo spazio della comunicazione.

L'eterno ritorno: dal mito ai social media (marketing narrativo)

Il mito non è mai una questione di denominazione e di nomi propri, ma di effetti che li sottendono. Dioniso e i protagonisti del mito sono, per i giovani e i comunicatori, dei "ferrivecchi", ma non lo sono gli elementi originari che agiscono alle loro spalle: è il percorso che Dioniso compie sulla via dell'emancipazione, come fanno i giovani.

Il percorso dal vino come bene originario, che scompare nella trasgressione e che riemerge come scelta consapevole, cos'è se non il percorso del dio che si immerge con la natura negli inferi per riemergere in primavera con il ritorno della vita? E Dioniso non è forse il dio dell'emancipazione e della liberazione in un cammino che viene ripercorso anche oggi? Per usare questo mito è, però, necessario ripartire da alcuni concetti implicati nel mito originario del vino: l'identificazione della dea terra con la madre vite, la dea vite da cui nasce la coppia origine/terra; la nascita del vino dall'unione di umido e secco, di aria e terra, di anima e spirito, di saggezza e passione.

Gli archetipi da evocare sono saggezza e passione per un consumo equilibrato (ambivalenza della saggezza e della passione del mondo greco); vino come marcatore culturale; conoscenza e gioia per un consumo maturo; conoscenza di sé e degli altri per una relazione migliore che regala piacere (vino, sangue di Dioniso, simposio); disinibizione/controllo; territorio e tradizione (tipicità, appartenenza).

Bibliografia selettiva

Acerbi, G. (1825). *Delle viti italiane, o sia, Materiali per servire alle classificazione, monografia e sinonimia, preceduti dal tentativo di una classificazione geoponica delle viti,* Milano: Silvestri

Bachelard, G. (1975). *Il diritto di sognare*, Bari: Edizioni Dedalo

Carmagnola, F. (2006). *Il consumi delle immagini. Estetica e beni simbolici nella fiction economy*, Milano: Bruno Mondadori

Cron, L. (2021). *Story or die. O racconti o sei fuori*, Palermo: Trefoglie

Del Zan F., Failla O., Scienza A. (2004). *La vite e l'uomo, dal rompicapo delle origini al salvataggio delle reliquie.* Gorizia: ERSA

De Martini, A. (2017). *Brand Narrative Strategy. Il segreto dell'onda*, Milano: Franco Angeli

Donati, Marco (2006). *Aristotele Retorica*, Milano: Mondadori

Floch, Jean-Marie (2002). *Semiotica, marketing e comunicazione,* Milano: Franco Angeli

Fontana, A. (2014). *Manuale di storytelling.* Milano: Rizzoli ETAS

Foucault, M. (2019). *Discorso e verità nella Grecia antica*, Roma: Piccola Biblioteca Donzelli

Giardina, A. (2011). *Il marchio demiurgo*, Milano: Lupetti

Goleman, D. (1995). *Emotional Intelligence*, New York: Bantam Books

Guyot, J. (1868). *Étude des vignoblesde France pour servir à l'enseignement mutuel de la viticulture et de la vinification françaises*, Paris: Imprimerie Nationale

Heidegger, M. (1976). *Saggi e discorsi*, Milano: Mursia

Hobsbawm E. J., Ranger T. (1987). *L'invenzione della tradizione*, Torino: Einaudi

Incisa della Rocchetta, M. (1984). *La terra è viva*, Firenze: Libreria Editrice

Lehrer, J. (2017). *Immagina. Come nasce la creatività*, Torino: Codice

Levadoux, L. (1956). *Les popolations sauvages et cultivées de Vitis vinifera l.*, Paris: Institut national de la recherche agronomique

Meyer, M. (1997). *La retorica*, Milano: Il Mulino

Mondini, S. (1903). *I vitigni stranieri da vino coltivati in Italia*, Firenze: Barbera

Morelli, U. (2020). *I paesaggi della nostra vita*, Milano: Silvana Editoria

Nomi, E. (2001). *Il frutteto di famiglia*, Milano: Mondadori

Nussabaumer Knaflic, C. (2016). *Data storytelling. Generare valore dalla rappresentazione delle Informazioni*, Milano: Apogeo

Proust, M. (2020). *Alla ricerca del tempo perduto*, Milano: Oscar Mondadori Baobab

Robertson, R. (2000). *Globalizzazione: Teoria Sociale e Cultura Globale*, London: Sage Publications

Rizzolati, G. e Gnoli, A. (2016). *In te mi specchio. Per una scienza dell'empatia*, Milano: Rizzoli

Salomon, C. (2008). *Storytelling. La fabbrica delle storie*, Roma: Fazi

Salomon, C. (2014). *La politica nell'era dello storytelling*, Roma: Fazi

Sassoon, J. (2012). *Web Storytelling. Costruire storie di marca nei social media*, Milano: FrancoAngeli

Scienza, A. (2014). *La storia della civiltà del vino in Europa raccontata attraverso il DNA dei suoi vitigni*, in "Le frontiere nascoste della cultura del vino" Atti SPEA, Annali S. Michele a cura di Forni, G., Kezich, G. e Scienza, A.

Scienza, A. (2014). *Venezia ed il vino: il tempo ritrovato in "Il vino nella storia di Venezia"*, Cittadella: Biblos

Scienza, A. (2015). *Emozioni, simboli e comunicazione del vino in "Il Soave: origine, stile e valori"*, Consorzio Tutela Vini Soave

Scienza, A. e Imazio, S. (2019). *Sangiovese, Lambrusco and other vine stories.* Verona: Positive Press

Semprini, A. (2006). *La marca postmoderna*, Milano: Franco Angeli

Tagliapietra, A. (2008). *La metafora dello specchio*, Torino: Bollati Boringhieri

Trupia, F. (2024). *Viva la retorica sempre*, Milano: Piemme

L'autore

An internationally renowned academic and agronomist, Attilio Scienza is a full professor at the University of Mil an and the author of over 350 published works, including at least 24 books. He has been responsible for numerous national research projects in the field of agronomy, physiology and genetics of the grapevine. Since June 2018 he has been the Chief Scientist of the Vinitaly International Academy and is the host of the popular podcast show, Everybody Needs a Bit of Scienza.

Attilio Scienza, accademico e agronomo di fama internazionale, è Professore Ordinario all'Università di Milano e autore di oltre 350 pubblicazioni, tra cui almeno 24 libri. Ha guidato numerosi progetti di ricerca nazionali nei settori dell'agronomia, della fisiologia e della genetica della vite. Da giugno 2018 ricopre il ruolo di Chief Scientist della Vinitaly International Academy ed è il conduttore della popolare serie "Everybody Needs a Bit of Scienza" di Italian Wine Podcast.

Altre risorse

Italian Wine Podcast

Italian Wine Podcast nasce come progetto di *storytelling* dedicato al mondo del vino italiano. Nasce nel marzo 2017 in occasione di due dei principali eventi vinicoli italiani – OperaWine e Vinitaly – ed è da allora diventato un programma pluripremiato, riconosciuto come uno dei podcast sul vino più ascoltati in circolazione. All'interno dei propri confini, l'Italia ha più vitigni autoctoni di qualsiasi altro Paese al mondo: la diversità del vino e delle uve italiane è una storia che vale la pena raccontare. Parallelamente, anche per accompagnare la crescente popolarità del vino italiano in tutto il mondo, i programmi si sono aperti ad ospitare anche personaggi della comunità vinicola internazionale. Italian Wine Podcast si propone di informare, educare e intrattenere i fan attraverso conversazioni coinvolgenti con produttori, esperti e personalità della scena vinicola italiana e internazionale.
italianwinepodcast.com

Vinitaly International Academy

Vinitaly International Academy (VIA) si propone come standard di riferimento per la formazione sul vino italiano. Offre un percorso formativo completo con corsi intensivi che insegnano a professionisti ed educatori del settore a padroneggiare la competenza sul vino italiano in modo rigoroso e organizzato. Maggiori informazioni sono disponibili su sito dell'istituto.
vinitaly.com/academy/vinitaly-international-academy/

Mamma Jumbo Shrimp

Il canale YouTube Mamma Jumbo Shrimp fa parte dell'omonima famiglia di contenuti multimediali legati al mondo del vino. Il canale si occupa anche di cibo e viaggi, con contenuti video regolarmente aggiornati e girati in tutto il mondo. Dalle interviste con iconici produttori di vino, incontrati presso le rispettive tenute, alle degustazioni di vino in abbinamento a piatti locali, ogni aspetto della cultura del vino italiano viene esplorato in video divertenti e coinvolgenti.

mammajumboshrimp.com

Della stessa casa editrice

Italian Wine Unplugged 2.0

Following on from its first edition published in 2017 (Italian Wine Unplugged - Grape by Grape), 2.0 is the result of a community-wide project: the best things need a village for the growth process. Covering all the Italian native grapes, seasoned with history and food pairings, the book includes substantial new material drawn from Professor Attilio Scienza's extensive groundbreaking research into the origins of Italian grapes and their DNA, giving the readers the opportunity to combine the study of Italian grapes with access to never-before-published material written by one of the leading lights in the academic study of Italian wine. This book is a must-read for students of Italian wine and wine lovers, providing up-to-date, useful and accessible content about a multi-layered and challenging subject. Included for the first time in this edition are benchmark producers for each grape, making it easier for wine enthusiasts to source and taste reliable, quality examples of Italian wines currently available on the international market.

Jumbo Shrimp Guide to Italian Wine

Talking about Italian wine is not always easy, especially if you are looking for a quick fix that can explain the complexities of this world. The Jumbo Shrimp Guide to Italian Wine was created to make the intricacies of Italian wine more approachable

for anyone interested in understanding it in detail but who may feel intimidated by the subject. This book offers all the tools to easily remember the key elements: how to read and understand the labels, the geographical information underlying each wine, the various types of grapes, and the most significant wines that could act as representatives of what is likely the most varied and qualitative wine production model in the world. It successfully delivers the most important building blocks for Italian wine education in an approachable way; the main producers, the great wines to watch out for, food and wine pairings, and last but not least a useful dictionary to distinguish between the places, grapes and the names of the various wines.

The Jumbo Shrimp Guide to International Grape Varieties in Italy

The book was created thanks to the collaboration between three writers who realized that in the current trend towards more and more obscure varieties, many great grapes, denominations and wines being overlooked in new Italian wine literature. While Italy boasts over 500 native grapes, some of its finest wines are made from international varieties. This book explores not only these wines but also their history in Italy, how they gained popularity, and how they are performing today. It also addresses what defines an "international" grape and how wines made from these varieties in Italy differ from those produced elsewhere. Designed for both casual consumers and wine students, this guide offers key insights in an approachable, fun format that can be explored in any order. With its characteristic style, the book serves as a valuable tool for both newcomers and experienced tasters alike, presenting information in a way that encourages readers to dip in and out at their leisure.

Wine Democracy

Wine Democracy presents the idea that wine should be accessible to all, regardless of income, background, or status. Through conversations with the hosts of the Italian Wine Podcast, this book delves into key challenges facing the industry today, including technology's influence, the exclusivity of "winespeak," and the ongoing struggles with discrimination and misogyny. Each host brings their own perspective to the topic, but all agree that greater empathy and awareness are necessary to make the wine world more inclusive. The book is filled with real-life experiences from those who live and breathe wine, as well as stories from people previously excluded from the industry. By highlighting the barriers they face and how they overcome them, Wine Democracy shines a light on the inequities of the sector while offering practical steps to create a more equitable wine world. It's an essential read for anyone interested in making wine more accessible and inclusive.

Vine and Prejudice - Fake Science and the search for the perfect grape

Is science prejudiced? What is fake science and how has it affected the world of wine? How can a deeper understanding of science help wine producers to confront the major challenges?

Professor Attilio Scienza takes a unique approach in this thoughtful history of the vine, tracing how grapes evolved, spread across the globe, and led to the variety of wines we enjoy today. From the earliest grape farmers to modern scientists, the book covers the milestones of vine science, including the development of cross-breeding and DNA profiling. Scienza also delves into the controversies surrounding science, examining fake news and ethical dilemmas. He asks readers to reflect on the future, particularly the role of genetic engineering and biotechnology, and what these advances mean for the future of the vine. Blending science with sociology, the book offers a thought-provoking look at how human intervention has shaped the vine's history and evolution, while also questioning where this journey will lead us next.

Sangiovese, Lambrusco, and Other Vine Stories

Behind the cultivation of grape vines and their fruits' fermented juice lies a history of peoples and global communities. Attilio Scienza and Serena Imazio take readers on a journey through the history and genetics of iconic Italian grape varieties like Sangiovese, Primitivo, Moscato, and more. Using DNA analysis, archaeology, and anthropology, the authors trace the ancestry of these grapes, exploring their origins and the cultural exchanges that shaped their evolution. The book also examines internationally known varieties such as Chardonnay and Merlot, along with the devastating phylloxera epidemic of the 19th century. Combining scientific insights with captivating stories of migration, conquest, and exploration, this book reveals the deep connections between human communities and the vines they cultivate. Written in an accessible yet scholarly style, it's perfect for both wine professionals and enthusiasts, offering readers a fascinating glimpse into the genetics and history of the vines that have shaped the world of Italian wine.

Printed in Italy by
Grafiche Mave
March 2025